ROBOTS BEHIND THE PLOW

ROBOTS
BEHIND
THE PLOW

Modern Farming and
the Need for an
Organic Alternative

by
Michael Allaby and
Floyd Allen

RODALE PRESS, INC./BOOK DIVISION
Emmaus, Pa.

International Standard Book Number   0-87857-077-2
Library of Congress Card Number 73-16812
Copyright © 1974 by Michael Allaby and Floyd Allen

First printing—February 1974

Printed in the United States of America on recycled paper.

CONTENTS

chapter I

THE SUGAR ON YOUR STRAWBERRIES

Within the next few years you may have to make a choice about the food you eat, a positive, conscious choice that will affect your diet and therefore your health. It will do more, however. It will affect the way the food you eat is grown and so it will affect our farms, the countryside, and the environment.

Have you eaten any "fresh" strawberries recently? (The " " are deliberate, as you will see in a moment.) Did you have to sweeten the strawberries with white sugar? Can you remember when fresh strawberries were so sweet and luscious that sugar would have interfered with their natural flavor?

This is the choice you must make, between sweet or sour strawberries.

Put it this way, of course, and the choice is simple. Who would choose a poor, deficient strawberry rather than one full of natural flavor and goodness? The choice may be obvious and simple to you, but it is less obvious and simple in market places, supermarkets, packing sheds, processing plants, truckers' terminals and the fields where the strawberries are grown. There the choice blurs, because while the two strawberries look very much alike, one of them, the sour one, was more profitable and easier to grow and to handle. Probably it was produced, harvested, packed, shipped and sold using methods that solve problems more conveniently and cheaply for everyone—except you, the consumer, the eater of strawberries. It may be, though, that the producer of sour strawberries is solving his problems and those of his colleagues in ways that will create far worse problems tomorrow, for everyone.

The choice, then, must be made by you, the consumer, but it must be made by farmers as well. If you choose organically grown food, you must become a diligent, skillful, well-informed food buyer. If you plan to grow it, you must become a careful, skillful, gentle farmer. If you do not choose organically grown food, you may forfeit your opportunity to choose. The option may close and in a few years' time the only way for you to obtain better food will be to grow it yourself.

It is quite possible that you live in one of the many areas where the option has closed already, at least so far as strawberries are concerned. For nearly 200 million North Americans, more than 50 million Britons, as well as countless more Frenchmen, Germans, Dutchmen, Belgians and Scandinavians, not to mention the Czechs Rumanians, Bulgarians and Poles, fresh strawberries look good, but taste sour. Frozen or canned strawberries taste sweet only because sugar is added to them when they are packed. They say that in Israel, anyone who knows anything about Israeli farming methods eats only home-grown strawberries, or tomatoes for that matter.

In the United States, 80 per cent of strawberry consumers now prefer sugar on their strawberries, which suggests that, without sugar, most strawberries are inedible. No one knows how many people have stopped buying "conventionally produced" strawberries, but the rapid growth of the organically grown food industry and the increasing popularity of home gardening suggest that large numbers of people are determined to find something better. The rate at which this is happening is phenomenal and very significant. Nor is it confined to housewives. There is a growing number of restaurants, hotels and institutions that serve organically grown food, and it is beginning to appear on university campuses.

A similar change in food buying habits is beginning to make itself felt in Britain and in Europe, while in some Asian countries the public dissatisfaction with the inferior flavor and reduced nutritive value of the new, high-yielding "miracle" rices has prompted governments to seek new varieties, other fertilizers and other farming methods that are capable of producing good flavor and nutritional quality, as well as high yields. The most successful of the "miracle" rices, IR8, cooks badly, has chalky spots and a protein content far below that of traditional varieties. One effect of its introduction has been to increase prices for the old, preferred varieties.

Even if the choice seems to have been lost, even if, where you live, it is impossible to buy organically grown food, it is not too late to make your voice heard and to begin to reverse a trend that is nutritionally, agriculturally and socially damaging. Your choice will be influenced by and will influence your attitude as a consumer, as a farmer, as a human being who needs to maintain his or her body in a state of good health, as an individual who wants to keep open as many opportunities as possible for everyone, and as a citizen of this planet who wishes to provide future generations with a stable, healthy, intact world.

If we fail to understand the consequences of avoiding this choice, current trends may continue. If they do, we will find that our food is produced by mammoth industrialized enterprises owned or controlled by a very few major corporations. The United States could find itself with fewer than fifty conglomerates; and Britain, as part of an integrated European agricultural industry, could find its food produced by multinational corporations. Even today, in Austria, poultry comes from Holland, eggs from Bulgaria. French companies have suggested that Normandy can keep most of the Common Market supplied with milk and butter and Britain plans to permit the dumping of EEC butter surpluses on her home market. It can happen and it will unless firm action is taken to prevent it.

In Washington, Brussels and London it is official policy that food production should be handed over to the major industrialists to be run like a chain of automobile factories. The US Department of Agriculture was founded, in part, to help small farmers, but in 1972 it revealed its long range plans for mechanized farms where single "plots" cover entire counties.[1] Already, in California, it is possible to drive at 65 miles per hour along a straight road for more than 35 minutes, passing continuous orchards on both sides, all owned and operated by a single system.[2]

Perhaps we should expect this. We live in the aftermath of the industrial revolution, which has changed every other form of production. For many years agriculture has escaped the attention of the industrialists. The returns on capital invested were inadequate and capital drained away from agriculture rather than to it. Now, however, pressure on land resources coupled with inflation in all the industrial nations has made the ownership of land attractive. Bankers, speculators and industrialists are buying land as an insurance against further inflation and, having bought it, are turning their attention to farming. They have been brought up, educated and profitably employed all their lives in industry and commerce and it is not surprising that they approach farming as they would any of the enterprises with which they are more familiar. One of Britain's largest poultry producers is an engineer by training and he has pioneered factory methods of egg and broiler production. A young millionaire real estate tycoon bought 10,000 acres of estates in Hampshire, England, late in 1972. It includes an entire village which he now owns outright. His intention was to protect himself against inflation. From time to time he visits his estate by helicopter and spends an hour or so touring it.[3]

It is not surprising, but it is highly unnatural. We may be seduced by our apparent technological expertise into believing that our farmlands can become a factory floor on which automated equipment will produce an endless supply of food in standard shapes, sizes and colors. We may argue that if cars can be mass-produced, then why not strawberries? If and when this happens we are deluding ourselves in the short term and in the long term we may be doing damage that will be very difficult to repair. Plants are not machines and they respond in different ways.

Yet, for the moment, a fairly close approximation to strawberries can be mass-produced. The system that mass-produces them, however, creates problems that can be solved only by

The tomato is one of the biggest victims of industrialized agriculture. Few farmers choose varieties for flavor or nutrition anymore. Rather, they choose a tomato that will be easiest for them to grow, harvest, and market. This means ones that mature all at the same time so that huge tomato harvesters can cut the entire crop down in one fell swoop; ones that are regularly shaped so they fit neatly into plastic containers; and ones that have thick, hard skins that don't bruise during the many miles they travel to reach distant supermarkets. Today, about the only way we can get really good-tasting tomatoes that meet our standards—and not industry's standards—is to buy from an organic farmer or grow them in our own gardens.

making sacrifices. In this case it is the flavor that has to go. You may feel that such a solution is unsatisfactory, and you are right. You may feel that it is not so much a solution as a transfer of the problem to another area, and you are right.

On the other hand, perhaps you like sugar and can see nothing wrong in adding it to your strawberries? If so, you are missing the point. If your strawberries are sour, it is because natural sugars that should be present are missing. The fruit is deficient. The sugar, or the lack of it, is easy to detect, but if sugars are missing, what else may be missing that you cannot taste? Is it possible that vitamins are absent, too, or other nutrients—some of which have not yet been identified? It is more than possible: it is probable.

Even if we confine ourselves to the loss of sugar and assume, for the moment, that this is all that is missing, you are being cheated because you must pay to put back a flavor and a nutrient you have every right to expect to be present in the fruit you buy. You are being damaged, too, because the refined sugar you return is known to be harmful to health. Extensive epidemiological investigations have shown that a change from a diet of organically grown food, from which nutrients have not been lost, to a diet in which large amounts of refined carbohydrates—sugar, white flour and polished rice—are consumed in a highly concentrated form is associated with a wide range of diseases, including peptic ulcer, diabetes, coronary thrombosis, appendicitis, diverticulosis, hemorrhoids, femoral thrombosis, obesity and varicose veins.[4] We eat far more sugar than we should partly because the taste is pleasant but partly because the natural sugars in our diet have been reduced and we overcompensate because with such a concentrated product it is difficult not to overcompensate. So sugar is habituating and our palate is deceived.

If we did not add sugar to our strawberries they would be unpalatable. Perhaps we should be guided more by our palate. We have a sense of taste in order that we may identify those foods that are nutritious and reject those with little value and substances that are actually poisonous. Food that is nutritious tastes pleasant and so we enjoy eating it. Where all flavors are

natural we may be guided by our palate into choosing a satisfactory diet, but when flavorings deceive the palate, nutritional
imbalances and deficiencies may result. Sour strawberries cheat
you by depriving you of the pleasure you are entitled to expect
from eating them.

You can detect a sugar deficiency, but others may be more
difficult. Deficient tomatoes, for example, are simply "flat"
tasting, or watery and lacking texture. Probably, such tomatoes
were picked several weeks before you bought them, while they
were still fairly dark green. They were held in cold storage and
then gassed to give them color. Wateriness generally indicates
forced growing in an unbalanced soil and heavy applications of
artificial fertilizers. Lady Eve Balfour, one of the founders of The
Soil Association which pioneered organic farming, once defined
the successful use of artificial fertilizers as "the art of making
water stand up," and it has gone so far now that some California
canneries are reported to be testing tomatoes for their water content, which the industry estimates was 10 per cent above average
in 1972, and deducting the excess weight before paying the
producers. Unfortunately, no one bothers to weigh the water
in "fresh" tomatoes. You pay for the water, and for its transport,
when you buy them.

A significant part of the increase in yields achieved by modern
farming methods is due to the increased water content of crops.
We could produce endless lists of examples of fruits, vegetables
and meats whose water content has been increased by modern
farming. Heavily fertilized potatoes, to cite but one, contain as
much as 17.5 per cent more water than potatoes grown with natural fertilizers,[5] and the trucks carrying them actually stream water
as they leave the fields. Does this explain why your potatoes taste
of nothing and do not cook well? Do they go black when you boil
them? This is due to excessive use of nitrogen fertilizers. So long
as we are prepared to buy water at the same price as food, there
will be an incentive to go on increasing the water content. Some
fruits are actually treated with stimulants to increase their water-
holding capacity, and progressive scientists have discovered a
cheap (cheaper than meat, that is) chemical additive that can be

added, with water, to chickens after slaughter in order to increase their weight.[6]

Perhaps you do not mind buying water at an inflated price? Again, this misses the point. Weight for weight, or bulk for bulk, since it is bulk that determines how much you are able to eat, the food must be deficient, simply because it is, literally, diluted.

There is a growing mountain of evidence from the United States and from the most intensively farmed countries of Europe to suggest that soils and the health of animals are suffering from the kind of industrialized farming we have practiced for the last twenty to thirty years. At the same time, the incidence of degenerative diseases increases each year and it is widely accepted by medical researchers that many of these diseases are caused, at least partly, by poor diet. Even overt malnutrition is making a small come-back in some areas. It is all but certain that the nutritional quality of modern fresh food has declined, and it is quite certain that most processed foods are deficient.

It is difficult to determine precisely how much blame rests with badly designed diets and how much with inferior food, but there are very few nutritionists today who will place any blame at all on the condition of the soils in which crops have been grown, although we know that some soils are deficient. Scientists, as well as the public, are misled by reductionist methodology, the belief that it is possible to explain puzzling phenomena or complex organisms as "nothing but———." Much of the organic movement and much of this book exists as a challenge to, and an indictment of, reductionism.[7] Scientific methodology deals only with what can be determined by its own methods and ignores, often as "irrelevant," dimensions of human experience and awkward data that do not fit neatly within its confines. The "nothing but" approach leads us to generalize. If spinach consists of "nothing but" so much of this and so much of that, we go on to suppose that all spinach plants are identical. It is true that crop plants are more uniform genetically than are, say, human beings, so that to regard all the individual plants of a particular species as identical is less untrue than if we were to regard all Americans, or all Englishmen, as genetically identical, but there is a world of difference between "less untrue" and "true." In fact, all plants vary considerably, even though they may have been grown in similar soils, in similar climates and by similar methods.[8] Two spinach plants growing side by side in a row of spinach may differ greatly from one another. Even leaves on the same plant differ. A cabbage leaf on the outside of the plant receives far more sunlight, warmth and rain than do leaves close to the heart. Their rate of photosynthesis is different, their uptake of atmospheric nutrient is different and, consequently, their chemical composition, texture, color and flavor are different. So when they tell you a cabbage contains nothing but——, what do they mean? You could be taking home a bag of chlorophyll, roughage and water.

Some nutrients are lost from fruit and vegetables very quickly after harvesting, yet produce may be days, or even weeks old when it reaches the stores. It can be picked before it matures and then treated with gases and dyes to "revive" it after it has been kept in storage. It has been carried, possibly thousands of miles, and then held until market prices rose sufficiently to ensure the maximum profit. As nutrients are lost, so are flavor and texture, so that by the time it is eaten it may be little more than a shadow of a product that was, perhaps, of dubious quality to begin with.

We have talked of the nutrients that are lost from food, but what of the substances that are added? No one knows, even approximately, the cost to our health services of the subacute, chronic poisoning that may be caused by pesticide residues and food additives, but the residues are there. Many of the chemicals used in agriculture and food processing have been very inadequately tested for their effects on human health, but such evidence as does appear is disquieting, and the rate at which new additives are introduced and withdrawn from lists of those permitted by the health authorities does not inspire confidence.[9]

Do we have to eat deficient, or even unsafe, foods? Is the sour strawberry the only one there is? Can we devise no way of producing our food that does not pollute the environment, abuse animals, damage the land, disfigure the landscape, reduce opportunities for individual enterprise, displace millions of human beings and replace farmers with computerized conglomerates which control the whole production-processing-marketing operation and threaten to establish a new feudalism, less socially responsible than the feudalism we believed we had overthrown?

The answer is that an alternative does exist, and has always existed, but before we can avail ourselves of it on a national, or international, scale, we must understand that today food is being produced and distributed with the wrong attitudes, for the wrong attitudes and for the wrong reasons.

If we are to understand these attitudes and reasons, we must first agree about why we produce food. Unless we establish this quite clearly, we will find ourselves confused by rationalizations that lead us to compromise by accepting as necessary technologies that are no more than convenient, and as indispensable methods that are no more than profitable. Agreement should not be hard to find, because food has only two functions: to sustain the consumer in a state of good health and to give pleasure. We identify good food by its flavor, as we have seen, so nutritious food must taste good. Thus the first aim of farming must be to produce food for good flavor and maximum nutritive content. The farmer has a second, and no less important, obligation. He must conserve the soil so that future generations may be able to continue to grow on it food as good, and as tasty, as the food we enjoy. All the techniques of agriculture should be directed to the production of tasty, nutritious food, while maintaining the fertility of the land.

Speaking at a conference organized by The Soil Association in 1969, Sir Joseph Hutchinson FRS, emeritus Drapers' Professor of Agriculture at the University of Cambridge and one of Britain's most eminent agricultural scientists, said that for a quarter of a century we had been performing a vast experiment with our agriculture, the results of which were just beginning to appear. He was talking about the effects of modern farming on the soil and he implied that modifications were necessary if a deteriorating situation were to be arrested. In fact, the "experiment" has been going on for forty years or so. Scientists have shown that when we depart from the basic principles of sound husbandry our farming has an adverse effect on the environment, the fertility of our soils is jeopardized and we become dependent for our food on materials, such as fossil fuels and certain minerals, that one

INGREDIENTS: ROLLED OATS, WHEAT GERM, HONEY, CASHEWS, SUNFLOWER SEEDS, PEANUTS, RAISINS, SOY OIL, COCONUT, SESAME SEEDS.

THE GOOD SHEPHERD CEREAL CO.
A DIVISION OF SOVEX, INC.
COLLEGEDALE, TENN. 37315 U.S.A.

The food industry has led the American public to believe that today's food must be highly processed and loaded with food cosmetics and preservatives to be marketable. But an increasing number of consumers are discovering that they do have a choice between naturally prepared, really nutritious foods and the overchemicalized commercial food that has become so much a part of American life. Breakfast cereals is just one area where shoppers do have a choice. Granolas and other wholegrain cereals are giving sugar-coated candy cereals real competition.

INGREDIENTS

Sugar, oat flour, degermed yellow corn meal, wheat and corn starch, dextrose, salt, corn syrup, coconut and peanut oils, gelatin, calcium carbonate, monoglycerides, sodium phosphate, sodium ascorbate, artificial flavors, artificial colors, niacin, iron, gum acacia, citric acid, vitamin A palmitate, pyridoxine (vitamin B_6), riboflavin, thiamin, vitamin D and vitamin B_{12}. BHT added to pre-

day will be exhausted. The conclusion is so obvious that you might be forgiven for wondering why the experiment was considered necessary in the first place!

We should agree, then, that food must be produced using methods and materials we know will produce the qualities we desire, that will protect the environment, and that will conserve the soil, maintaining or enhancing its fertility. Can we afford to abandon or even to compromise any of these principles?

Thus far everyone agrees, or claims to agree, but then the rationalizations begin. "Is fertile soil so important when all that plants require for growth is (nothing but) potassium, phosphorus and nitrogen, which we can supply to them cheaply and precisely from a bag?" Yes, a fertile soil is important. This grossly over-simplified view of plant nutrition has done untold damage. We still do not know just how many nutrients plants do require. There are 16 chemical elements known to be necessary and these are provided by the soil in, at the latest count, at least 60 different compounds. To make them available to plants there are vast networks of bacteria and other soil microorganisms that are still only poorly understood. The more soil ecologists learn, the more convinced they become that the soil is one of the most complex living systems on the planet, with thousands of different species of microscopic plant and animal, each playing its own specialized role. Certain humic compounds, for example, are of great importance to the healthy growth of plants. The availability of humic compounds depends on the presence in the soil of adequate quantities of humus and organic matter, together with healthy, balanced populations of organisms. These in turn utilize, store and make available essential trace minerals. Humic acid is known to facilitate plant uptake of zinc. Superphosphate, on the other hand, a chemical fertilizer, inhibits zinc uptake.

It is true that nitrogen (N), phosphorus (P) and potassium (K) fertilizers promote growth in a fertile, balanced soil. It is also known that "NPK" farming depletes the soil, reduces the humus content, unbalances nutrients in the soil, affects micropopulations adversely and, as we have seen, produces deficient food and food with excess water. Yet NPK farming continues. Most of the food produced in Europe, North America, Japan, Australia and New Zealand, and an increasing proportion of that produced in the nonindustrialized countries of Latin America, the Near East, Africa and the Far East are grown with NPK fertilizers, hybrid seeds, synthetic pesticides, extended irrigation and bigger and bigger machines. In a way, this is what modern "scientific" farming is all about.

So far as the nonindustrialized countries are concerned, the immediate effect is to cause internal social and economic disruption and to tie their economies ever more closely into a subservient position to that of the industrial nations. Aid often takes the form of investment for profit and the "Green Revolution" is good business for the agroindustrialists of the rich countries.

Yet many of us believe that we are helping them and we are quite sincere. We have convinced ourselves that this is the cheapest way to produce the largest quantities, which is the aim of most modern farming. Look at any large-circulation farming magazine in any industrial country and the theme of all the articles and the advertisements is the minimization of costs and the maximization of profit. This is sound enough, up to a point, but the techniques have failed, as we shall show later, and the costs are kept artificially low.

No one really cares about the nutritive value of the food.[10] When the supply of nutrients to plants or animals is discussed, the object always is to increase yield or the rate of growth. Ruminant animals are fed diets unnaturally high in protein, which is not good for them, and there is a good deal of research aimed at developing cereal varieties with a higher protein content. The object is not to improve the quality of the product, but only the speed with which the animal grows to the minimum size acceptable for slaughter. Animals have become "protein factories" and in addition to their overenriched diets they are fed growth-promoting hormones, such as diethylstilbestrol (DES), traces of antibiotics, "hot feeds," nitrogen, and anything else that may help increase their productivity. The other problem livestock farmers face is that of keeping their stock alive long enough to be slaughtered.

In an age dominated by antiquated economic theory, farm livestock can be bought and sold and so have a value. Human beings do not.

If you were to suggest to such a modern agriculturist, bemused by his own cleverness, that it might be preferable to concentrate rather more on quality and rather less on quantity, he would be profoundly shocked. He would tell you that food prices would rocket upwards as supplies dwindled. He would remind you of the world's starving millions, all dependent on him and his feed lot. You would see him bow with the weight of the responsibility he bears. He might even quote the Secretary of Agriculture, Earl L. Butz, who said: "Those who. . .advocate the hasty abandonment of agricultural chemical pesticides, herbicides, and nutrients should take another look at the realities of the situation. . . .We can no more afford to revert to farming methods of 50 years ago than we can go back to horse and buggy transportation or oil lamps and candles. . . .We can go back to those times and those methods if we have to—but if so, who will decide which 50 million or so of the population will have to go hungry?"[11]

In fact, he is worrying about the wrong problem. In America, and even in Britain, the world's largest food importer, as well as in the EEC countries as a whole, up until now the problems have been of surpluses, of overproduction rather than shortages.[12] Governments have sought ways to restrict production, by complex support systems or, as in the United States, by encouraging farmers to take land out of production. It is gluts that economists fear, not famines, while farmers agitate to be allowed to produce to what they believe is their full capacity. Europe is seeking ways of disposing of its "mountains" of surplus produce. The need to limit production conflicts with the desire of the agricultural scientists to increase it. The clash of interests can lead to amusing paradoxes. In the spring of 1973, a spokesman for a leading research laboratory said, of poultry production, that "there is no problem of overproduction, only of under-consumption!"

It is true, of course, that one-third of the world population is chronically short of food, but the countries that are more than self-sufficient do not supply this deficiency. Production methods

in the industrial nations are far too costly, being based on plentiful supplies of energy, which are available cheaply to the rich but
not to the poor. In any case, lack of foreign exchange prevents
poor countries from buying as freely as they need to do on the
world market. Nor does the free supply of food from rich to poor
assist the situation except in the very short term to deal with
emergencies. The effect of US food aid in the 1960s, especially
to India, was to depress food prices in India, which was a good
thing, but hence to bankrupt Indian farmers, which was a bad
thing and led to a position worse in the end than it had been in
the beginning.[13] So we sell no battery eggs or broiler chickens to
India. No, our problems are with surpluses. Thus, a move to
organic methods of husbandry within the present context would
be likely to produce no shortages of food and in the United States
it would bring back into production land, communities and farming families made redundant by large-scale chemical farming.

What about prices? Presumably farmers use the cheapest
methods available to them. It follows, therefore, that any alternative is likely to be more expensive, otherwise it would be in use
already. So on the face of it we would expect food prices to rise.
However, the position is not so simple. The price you pay for the
food you buy in your store can be compared to the price that food
brings in world markets. In all the industrial countries of Europe
and North America, consumers pay well above world prices for
most commodities,[14] while we subsidize our food exports to keep
them below world prices. Our agriculture is extremely expensive.
As the drive to increase production per acre from larger and larger
farms has accelerated, farmers have been paid too little and consumers have paid too much. The chemical companies, the processors and the supermarket chains have grown bigger and richer,
but farmers have grown fewer in number. A reversal of this trend
would lead to a larger number of farmers farming more acres,
and a larger number of shops selling a wider variety of locally
grown produce. There would be fewer and smaller chemical companies, processors and supermarkets. It does not follow from this
that prices would increase. Indeed, there could be significant
savings in marketing and distribution.

The threat of increasing food prices is an emotional one and it is one feared by politicians. Yet it need not be. Its effects can be ameliorated either by price subsidies to the consumer or by wage increases. All the industrial nations experienced sharp increases in meat prices during the winter of 1972/73 because of world food shortages, yet in spite of vociferous protests, they survived. In Britain housewives have had to accept price increases in most food items as British agricultural policies were brought into line with those in the EEC and the old "cheap food" policies were abandoned. Yet they survived. In a rich country an increase in food prices means no more than that a slightly higher proportion of the family income is spent on food than was spent previously. If we object to this, we should remember the kind of world in which we live. Millions die each year from starvation or from diseases associated with malnutrition, and millions more watch their children maimed for life for want of sufficient quantity of nutritious food for them. If, within our own countries, the poorer sections of the community suffer hardship when prices of essentials increase, then we would do better to attack the problems of poverty than the system of food production. It is unfair to blame farmers for social problems and it is inappropriate and dangerous, for in blaming the farmers we may be seeking scapegoats in order to avoid the more inconvenient implications of living in an unjust society.

Mass production of food does not mean more food. It means food produced and distributed by a smaller workforce and fewer, larger concerns. Once there were nearly 6,500,000 farms in the United States. Soon there will be fewer than 2,000,000 and it may not be long, unless the trend is reversed, before some 200,-000 farmers produce food for 50 to 100 huge companies. This situation is being repeated throughout Europe. In Britain, the official figure for the agricultural workforce amounts to 3 per cent of the working population. Almost certainly this figure is too high and the true proportion is under 2 per cent.[15] The number is still decreasing and it is being encouraged to do so. At the same time in England, Wales and Scotland, some 60,000 acres are lost each year to urban and other nonagricultural development. As the workforce is reduced, so is the number of farms. A spokesman for a company owning several large estates and which is, in fact, a conglomerate of more than 20 building and real estate companies, has said on television that the days of the "stick and dog" farmer are over. He meant by this that there is no future for the farmer who has a close, intimate relationship with the land he works—in other words, for the small, family farmer. In his view, no farm should be less than 700 acres in size. In Britain this is usually considered a fairly large farm. Of a total of some 300,000 holdings, there are only about 5,000 of more than 500 acres. A tremendous number of family farms will have to go if this is to be the pattern of farming in the future. The United States loses 200,-000 farms a year as the same philosophy is applied. In the short term, this preoccupation with size may be profitable for some, but does it lead to attitudes we would choose to have associated with the production of our food? Does it lead to the production of sweet strawberries? We have seen, and your own taste buds have felt, that it does not.

chapter II

IT TASTES FUNNY, BUT IT FITS THE BOX

A major lettuce producer in Salinas Valley, California, who is a member of a large Western association which represents mainly industrialized agriculturists, summed up industrial farming very succinctly in a timeless statement of his philosophy: "I don't care what's on it," he said, "or what's in it, or what it tastes like, so long as it's the right size and shape and the right number will pack in a box and the box will pass the inspector." That man is a realist. He knows that size and shape can be sold, but that flavor and nutritive value cannot. He will welcome any product or technique that will help him mold his lettuces, make them conform to what a sensible, economic lettuce should be. He will buy new chemicals, new machines, new genetically engineered varieties and anything else that will give him more heads per acre of the right size at the right time. Quality standards have become cosmetic standards.

Some years ago Florida introduced "quality" controls on the movement of citrus fruits out of the state. The controls related solely to appearance and size, and in order to conform to them growers had to spray their fruit. An emeritus professor of moral philosophy from Yale University had bought his citrus fruits by mail from an organic grower for many years. He was outraged when he received a letter from his farming friend telling him he could have no more organically grown fruit. He composed a 23 page letter to the Florida Citrus Commission in which, in his professional capacity, he attacked their concept of "quality." He won his fight, but it is not every grower who can call on such a formidable ally.

If you insist on buying food that has been grown organically, you will join the ranks of those who are prepared to pay a little more for maturity, flavor and better nutrition. You will begin to care about how your food was produced and how long ago it was harvested. You will not be satisfied with an apple whose inside is green, no matter how red its skin. You will not buy more than once from the grower of such apples. You will not buy a second time from a grower whose apples begin to show bruise marks after they have been in your house for several days. Such apples were picked unripe. Unripe apples do not show bruises below a certain temperature, but as they warm to your room temperature the bruises begin to appear. You will not want deciduous fruits, such as peaches and apricots, that shrivel, dehydrate and become sour and rubbery after a few days. Such fruits were picked while they were still green.

You will find yourself buying from the smaller producers. The large-scale growers do not harvest several times, selecting those fruits that are ripe. They average out ripeness and take their harvest in one, or sometimes two, operations. In this way they can manage their large orchards and vineyards. Yet if fruit is to be picked in the peak of its condition, several harvestings are unavoidable. Only the smaller growers are able to provide the detailed attention necessary for four or maybe five harvestings, and they will aim to sell locally, because the fruit is picked ripe. Unripe fruit is easier to pick. It can be refrigerated more easily and it travels well, which means it can be held in storage until market prices are right for its sale. Buyers, who purchase in bulk for supermarket chains, can specify cosmetic standards by the carload with little fear of complaints from store managers. Prolonged shelf life enables prices to be adjusted. Large-scale producers believe no one will pay them for the maturity of their crop and it is certain that they will incur the whole of the risk involved in leaving the crop on the trees until it is mature. Nor can they sell locally. Their output is so great that it exceeds by far any possible local demand.

Of course, this is something of a generalization. There are exceptions. Some growers take great pains to produce tasty, nutritious foods and seek to develop storage and distribution techniques that conserve the quality of their crops. Similarly, there are packers who care about the quality of the food that reaches your kitchen and retailers who will pay extra for better quality goods and who handle food well. Such conscientious farmers, packers and retailers are fighting low-grade mediocrity and greedy, indifferent compromise.

For the good farmer, poor food is bad business. People are buying fewer fresh foods. In the United States it has been estimated that volume sales of fresh produce are 25 per cent lower than they were prior to World War II. This trend is attributed to the move to "convenience" foods—frozen, prepacked foods. There has been such a move, of course, in all the industrial countries. It is associated with increases in the number of young wives in full-time employment and as such it is part of a wider social phenomenon. At the same time, however, there has also been an increase in the amount of food sold directly by the grower to the consumer. This, too, is part of the same social change, as more people have cars and use them to visit the countryside. Farmers have welcomed it and exploited it. As the harvesting of fruit becomes more and more expensive, growers encourage the public to call and pick their own. There is an increasing number of roadside stalls and in parts of England farmers' organizations are encouraging their members to expand this profitable trade as a protection against their large-scale competitors and the processing companies. In the United States this form of trading has increased to such an extent that several major universities have conducted surveys, all of which show that more Americans are driving farther to buy more, fresher, and better quality food. Nor is it only fruit and vegetables that are sold in this way. Fresh, free-range eggs are advertised every few miles along thousands of miles of British country roads and as often as not poultry is for sale as well. Whether they are genuinely free-range is controversial, but even if the producers are cheating, at least they recognize that it is for free-range produce that the market exists. Honey, flowers and pot plants are frequently offered and in the United States it is now possible to buy meat in this way.

In a way, "buying from farmers" is what this book is all about, for whether you buy directly from the farm or from a shop, we believe you should buy with the farmer in mind and he should grow food with his customers in mind. You and he should agree about the kind of food you are to eat. Unless farmers and their customers can reestablish an understanding of each other's needs, neither of them will have much voice in the kind of food production and marketing that will come in the next few years. The farmer will grow for, and you will buy from, industrialists, bankers and speculators. Food will be more uniform, but it will not be better. It will be plentiful, but it will not be cheap.

There are farmers whose concern for the welfare of their consumers extends beyond the desire to do their job well. Most small farmers, and especially organic farmers who know they are growing for consumers who appreciate the quality and flavor of their food, are curious about their customers. They like to meet them, to get to know them, so that the relationship can become more than simply a business one, like that between the Florida citrus grower and his professor friend, or like that between Arthur Hollins, the British producer of "Fordhall" dairy products, who receives up to 5,000 visitors a year at his farm and who spends many of his winter evenings lecturing and meeting people—consumers. He is selling his produce, of course, but he is doing more than that.

It is the small farmer who is under pressure. The US Department of Agriculture was reorganized in 1933 under the terms of the Agricultural Adjustment Act. Its task was "to help small farmers." But the small farmers have been sold out. In the almost fanatical belief that big is best, every encouragement has been given to increasing the scale of each stage in the production and selling of food. The USDA has been seduced by the agrochemical and machinery industry.

Consider the impetuosity with which our agroindustrialists rush to buy and spray modern pesticides, such as DDT. True, when they were first introduced we were all assured that these substances were entirely safe. We were told that they would end for all time the scourges of insect-borne disease and that crop pests would

Consumers and farmers are in a much better position to influence the quality of food produced where they understand each other's needs. By buying locally grown food directly from the farmer, consumers have a better idea where their food dollars are going and can understand more clearly how their food is produced. Growers who have personal contact with shoppers are better able to provide the kind of food that customers want.

By buying from supermarkets, however, consumers are supporting the lower quality, excessively wrapped food and impersonalization that go hand in hand with industrialized food production and marketing.

disappear, but these assurances were given on the basis of totally inadequate testing. Even today no one knows for certain what are the effects on the health of human beings of small doses ingested regularly over many years. If, as is possible, they have mutagenic or teratogenic effects, the results may not become apparent until the second or subsequent generations, by which time it will be too late to take remedial action. Experiments on small animals have produced disquieting results, which could have been produced thirty years ago.

Insects acquire resistance to insecticides, thus rendering them ineffectual in the long term. This is common knowledge among entomologists and has been for many years, but pesticides were developed first by chemists and there was little communication between the disciplines and too much secrecy as manufacturers sought to protect the research and development costs bound up in their products from piratical competitors.

Today, wildlife has been reduced, the oceans suffer from levels of pesticide contamination that may be serious, and food has been contaminated for a generation. Meanwhile the pests flourish. California has twice as many mosquitoes as it had in 1942, when water management programs were scrapped in favor of DDT. Throughout the tropics disease eradication programs are faltering as the insect vectors acquire resistance.[1] Pesticides are failing and now everyone agrees that we need to develop alternative methods of pest control. The fact is that we always needed alternative methods, and we could and should have known it before pesticides were introduced. They are unsound in theory as well as in practice. Under the strain of modern farming, of which pesticide use is an integral part, some of the world's most productive soils are suffering from depletion, compaction and erosion. Soil pathogens increase, while yields decrease and food becomes flavorless and deficient. We need alternative methods of farming.

Fortunately, such an alternative exists. If we were to draw up a list of desirable reforms for agriculture, we would see that all of them tend to move farming in the direction of organic methods. In the light of present knowledge, then, organic farming is by far the most satisfactory alternative to present practices.

What is organic farming? Perhaps you imagine that an organic farmer is one who uses no artificial fertilizers or pesticides? It is true: he does not, and in Britain attempts to define organic farming have led the Soil Association to produce a "code of conduct" that is based largely on lists of permitted and forbidden materials and techniques. Yet it is far more than that. The organic farmer is interested in results more than in techniques and he is involved in a total alternative to food production, processing, distribution and marketing. Because of his concern for the welfare of the consumer he must concern himself with the whole of the operation that brings your food to you.

Now, perhaps, you are confused. It is easier to think of a closed, industrialized or integrated system of farming than the kind of open-ended, diffuse concepts associated with the organic alternative. This is inevitable. We live in a very complex world and although we may appear to comprehend it better when we simplify our concept of it, what we comprehend is, in fact, no more than our own mental model, which we have distorted. Farming is complex. The materials it uses—soils, plants and animals— are complex living organisms and communities of organisms and they cannot be made simpler. If we imagine that biological processes can be understood in the way industrial processes can, we delude ourselves. We have not repealed the laws of biology or ecology and they remain as complex as ever they were, in the real world. However, if we apply our simplistic views to our farming systems, we will abuse and damage the raw materials with which we work. We may destroy the very fabric of life.

The word "organic" has been defined as "acting as an instrument of nature or art to a certain destined function or end."[2] We cannot improve on that definition.

No living thing exists in isolation. Everything has an environment—if you like, an ecological context. This context consists of a network, a web of countless associations with other organisms, with mineral substances, with the climate. If the web is so arranged that conditions are favorable, the organisms will flourish. If it is unfavorable, they will fail. Organic farming and gardening seeks to establish and maintain conditions within which plants and animals may thrive. It does this by close observation and constant adjustment, for environments are four-dimensional— they change with time. They are also extremely localized and can and do vary quite widely from field to field and from one part of a field to another part of the same field. The more we understand these local environments the more subtlety we can use in our husbandry and the more successful we will be in encouraging useful plants that can sustain themselves in their natural surroundings with the minimum of outside aid or interference. Organic farming is not based on a return to some romantic, and quite mythical, "golden age" of arcadian bliss. It aims for gentler, more intelligent, more scientific methods.

Here, though, we should say something about "knowledge" and "understanding," lest we be trapped into a reductionism indistinguishable from that which has led us into the present impasse. There is no doubt that we need more objective data relating to agricultural systems and the relationships between organisms within them, but this information will be of little value if we seek to apply it in order to increase our "control" over nature. This would be to reduce it to what Rosjack calls "power knowledge"[3] and its application would alienate us still more, when fundamentally our predicament can be traced to our alienation from nature. It would be more likely to exacerbate than to reduce the problem. We need knowledge, but even more we need understanding, knowledge in depth. This may manifest itself as an intuitive feeling for nature, an identification between the farmer and his land, plants and stock, which will tend to lead him to make decisions which are to their benefit. The point is subtle, but important, and it is this kind of feeling that is so common among good farmers in general and organic farmers in particular. It arises from the attitude of the farmer to his work and from his motives and the "dog and stick" farmer who walks his fields has a better chance of acquiring this feeling than the farmer who sees his farm only from the seat of his tractor. The farmer who sees his farm on a computer print-out has no chance at all. All the knowledge he acquires alienates him from the living world he seeks to dominate. If the aim of the farmer is only to become rich, he may succeed—but at a high price. The acquisition of wealth is an unsatisfactory motive and if that sounds moral, it is meant to. The production of food is a profoundly moral business.

Yet, provided we are quite clear about why we need it, we do need knowledge in the sense of data. We do need improved techniques. Organic farming is developing constantly and already it is practiced successfully by many farmers and growers in many countries, but if society were to devote even a fraction of the resources to its improvement that have been devoted to an inadequate synthetic technology, it could move even more quickly and the day would be closer when we might introduce on a wide scale an agriculture that avoids pollution and that respects the quality of produce, the condition of the land and the welfare of those who work on the land. All our strawberries could be sweet.

Indeed, we may have no choice but to phase out current farming methods and substitute organic farming. Modern industrial farming is sustained and made to appear efficient and economic by very large inputs of energy and minerals. Certain of the mineral resources required for fertilizer production are nonrenewable[4] and already the United States is experiencing an energy crisis. It is probable that Western Europe will experience a similar crisis, perhaps by the mid-1980s. When farming systems are compared in terms of their utilization of energy, "efficiency" acquires a new meaning. It has been shown, for example, that the Chinese small farmer growing wet rice is 6,000 times more efficient than the American farmer. This aspect of agriculture is important and we shall return to it later. For the moment, we need do no more than ask for how long we can continue to sustain a system which employs such inefficient technologies to provide short-term solutions to problems most of which it has created for itself?

The inspiration for organic farming came from several sources but especially, perhaps, from an Englishman, Sir Albert Howard. Born of farming stock, on the borders of England and Wales, he became a brilliant scholar. Indeed, his education was prolonged as he collected diploma after diploma; and it was not until he was 26, in 1899, that he took up his first appointment. He travelled widely as a university lecturer and then as a research scientist.

He was a mycologist, specializing in fungal diseases of plants, and his observations led him to formulate theories of disease resistance. His most impressive work was done in India, where in 1905 he was appointed First Economic Botanist to the Government. As a result of his long study of Indian agriculture he developed a theory of "wholeness" that is still valid today. He wrote that "the main characteristic of nature's farming can. . .be summed up in a few words. Mother Earth never attempts to farm without livestock; she always raises mixed crops; great pains are taken to preserve the soil and to prevent erosion; the mixed vegetable and animal wastes are converted into humus; there is no waste; the processes of growth and the processes of decay balance one another; ample provision is made to maintain large reserves of fertility; the greatest care is taken to save the rainfall; both plants and animals are left to protect themselves against disease." He believed that a civilization may be measured by its treatment of its farm land. It follows that an attitude to food production that is based on such a complete concept of the interrelatedness of a multitude of factors must foster a high state of ecological awareness that, in turn, will lead to a profound respect for the planet and for its physical and biological systems upon which we depend for our survival and well-being. Without increasing the extent of our knowledge, this change in our attitude will deepen our understanding. We can begin to experience what, cerebrally, we know.

Thus, organic farming and gardening advocates the obvious. It suggests that we have smaller farms, localized production, more direct marketing and therefore shorter distances over which food must be transported. At each point it reduces waste and so aims for higher efficiency. Much of the home-grown food that we eat could be produced locally. There is no reason why you should have to buy fruit and vegetables that have been transported over a long distance when they could have been grown nearby and supplied to you fresh. This is especially true of fruit and vegetables and in areas where demand exceeds current production, techniques are already available to increase production. It may be that no more is required than an increase in the area under glass, or sheltered by plastic windbreaks. Where foods must be transported over long distances, or stored, an improvement in the initial quality of the produce together with more careful handling will enhance storage properties without compromising nutritive value.

When we begin to design programs to convert farms and farming areas to the organic alternative, we find there are other, incidental benefits. We are adopting an approach that is radically different from that which is enshrined in national agricultural policies. We are saying, instead of making farms larger, make them smaller; instead of employing fewer men and more chemicals and machines, employ fewer chemicals and machines and more men. Small rural communities revive and offer new opportunities for small-scale businessmen to serve them. Employment increases in rural areas and, with it, prosperity. Pressure is relieved on urban welfare services as welfare becomes the responsibility of small communities to their weaker members. It is obvious that if, say, thirty farmers and their families each live on and farm 100 acres, they will provide a better life and solve more problems for more people than one farmer and his family operating 3,000 acres. Yet, in the United States today, a 3,000 acre farm is considered small, and there is one company, Tenneco, which operates one million acres in California and Arizona.

The reintegration of livestock and arable farming would reduce the problems attendant on intensive livestock rearing, there would be less pollution from both livestock and arable farms, and it would be much easier to incorporate town wastes.

Of more fundamental significance, however, is the fact that more people become involved in agriculture as a cultural activity. Too many of us have forgotten that for most of our agricultural history farming was an activity performed by whole communities. Many of the operations were also religious ceremonies, in which everyone participated. The idea that farming should be left to a few specialists is very new, and already there is a growing demand, especially from the young, to reestablish farming communities, to become involved in the production of their food. Naturally, it is in organic farming, and only organic farming, that they are interested.

To many orthodox agricultural professors, especially in Britain, where they are particularly educationally retarded, organic farming seems impossible. Thus they conclude, and teach their students, that organic farming cannot exist. The food industry and its army of tame nutritionists maintain that the demand for organically grown food is a fad that will pass. They choose not to notice that there are many true organic farmers, and successful ones, and that while it is true that the demand for organically grown produce has been increasing over the last few years, there has always been a quality market that demanded such food from such farmers. Certain of the most expensive London stores have always sold organically grown food. They have not advertised the fact, or labelled the food as such, because it seemed unremarkable to them. Their customers would accept only the best and they would supply only the best. This was the best. It was as simple as that.

Nor may the professors have observed that sometimes organic farmers are conspicuous because they are rather more prosperous than their chemicalized neighbors and their farms are in better condition. On occasion this can lead to ill feeling. There have been cases in Europe of organic farms being watched, day and night, by rivals eager to discover the secret spray that had been used!

Toward the end of 1971, demand for organically grown food in the United States exceeded the supply. As shortages occur, it is inevitable that food will be sold as organic under false pretenses. This has happened and it causes considerable concern to the organizations whose task it is to safeguard the reputation of organic methods and to ensure that the consumer gets a fair deal. Steps are being taken to introduce certification schemes, with registers of accepted growers. Their produce can be guaranteed. In Britain, the Soil Association launched its program early in 1973 and tied it to a mark that can be used to identify authenticated produce.

In the United States, state and federal authorities refused to step in and protect consumers and so the task was left to the staff of the monthly magazine *Organic Gardening and Farming,* which had spent years advocating organic methods and helping those who practiced them. They began a certification program in California by inviting organic farmers to apply. The farmer was asked to submit a commitment questionnaire which described his methods and the materials he used. If his statement was accepted he would be visited by a representative of Rodale Press, publishers of the magazine, and a sampler from Agri-Science Laboratories. If analysis of the samples showed no residues of persistent pesticides and examination of his farm suggested that he was farming organically, he was accepted for certification. From then on his farm would be visited quarterly and samples of soil, water and plant tissues taken for analysis. The cost was borne by Rodale Press and the scheme was operated by the hard-working editors of *Organic Gardening and Farming*.

Certified organic farmers were given the privilege of using a mark which included their name, city and state address. Immediately the customer gained confidence. He could see, not just that the food he bought claimed to be organically grown, but by whom it had been grown and where. He could go and see for himself if he had a mind to do so. In the United States, laws do not exist to protect the consumer. This is not the case in Britain, where once a definition of organically grown produce is accepted by the courts, the Trade Descriptions Act provides all the protection that is necessary. This Act makes it an offense to misrepresent a product. Already prosecutions have been brought successfully against retailers selling as free-range eggs those that had been battery-produced.

In less than a year the Rodale program showed that organic farmers did exist and that they are certifiable. At the same time, the California pilot scheme indicated the need for similar programs elsewhere.

Organic farmers in the state of Washington already had a viable certification program and Rodale Press assisted them in extending it. The farmers agreed to form an association, which they subsequently named Northwest Organic Food Producers, to define the area they would cover, to agree that all organic farmers in the designated area who met the minimum requirements would be eligible to apply for certification, to use the Rodale commitment and procedures questionnaire, to include any organic farmer with one-third of an acre or more producing edible commodities on a regular basis and selling at least a portion regularly, to adhere to the traditional "open house" policy, and to maintain records of organic farming methods in respect of each farmer together with verification based on laboratory tests and farm inspection. The farmers would pay all the costs themselves and would have the right to use the Rodale seal.

By early in 1973, the NOFP Association had established the principle of self-regulation for organic farmers so successfully that the Director of Agriculture for the adjoining state of Oregon announced that his office may officially establish a similar program and the Director of Agriculture for the state of Washington has expressed interest in endorsing the NOFP program.

On February 24, 1973, the organic farmers in California who were members of the program in that state agreed to assume responsibility for its administration and to establish an association similar to the one in Washington, called California Certified Organic Farmers (CCOF), to supply out-of-season vegetables. The CCOF organization will extend membership to consumers, retailers, cooperatives, organic fertilizer companies, distributors and others interested in promoting the organic alternative and who have embarked on an ambitious program over a wide area. In April, 1973, OGF announced that it would award the use of the Rodale seal to organic farmers in any state who will organize and establish a program to guarantee their produce.

One of the more interesting findings of the Rodale program is that there are many farmers who would like to farm organically. They feel they lack the information necessary to enable them to change their methods, but many nonorganic farmers are organic gardeners. They know that food can and should be produced without the excessive use of chemicals and they are convinced that farmers could grow food with a better flavor and higher nutritive value. Other farmers are attracted to organic methods because the cost of agrochemicals reduces their profits. There are still more farmers who know how to farm organically and who would like to do so, but who are prevented from practicing their beliefs by state and federal codes that are so fine that the farmer risks having his crop condemned if he exceeds them by even a half per cent.

Fraud on the part of farmers usually falls into one of two categories. There are those farmers who do not believe that organic farming is possible under current economic pressures and, in the United States, regulations controlling cosmetic quality, yet who wish to obtain the premium price paid for organic produce. Their fraud is deliberate and direct. There are others who use materials and methods that are not acceptable within the strict definitions of organic farming, yet who believe they are farming organically. Their fraud is not deliberate; they are acting in ignorance. The evidence from the Rodale experience suggests that the supply of organically grown produce is smaller than it might be for want of adequate sources of information and because of the restrictive nature of "quality" controls that require the use of chemicals.

The increasing interest in organic farming has revived an old controversy over differences in flavor. Are there differences? It is impossible to say, since taste is highly subjective. Most of the people who buy organically grown food believe it tastes better, but this is a poor indication, since they buy it expecting it to have more flavor. Such few double-blind tests as there have been were inconclusive. Yet there are indications. All over the world, farmers have observed, quite independently that, given a free choice, animals will prefer food or pasture that has not been treated chemically. Some of the greatest chefs will use only food that has been grown organically, because they are convinced that it tastes better. Animals are not influenced by propaganda and skilled chefs should know their business.

There is evidence, too, to suggest that food grown organically is nutritionally superior. The evidence is diffuse, but it exists. On March 12, 1973, New Scientist reported on research at the University of Texas on a new method for evaluating food on the basis of its trophic, or constructional, value. It was found that certain high protein foods, such as canned tuna, canned chicken and fortified skim milk have low trophic values and that free-range eggs may be significantly superior to battery-produced eggs. Rats fed on free-range eggs grew to a weight of 350 gms in 12 weeks, whereas those fed on battery eggs reached only 315 gms in the same period. The researchers concluded that unless food has sufficient trophic value, it can never allow an animal to realize its full potential, even though it may sustain life.

Seeds of millet and wheat from a field fertilized with manure were found to contain more vitamins than seeds from a field treated with artificial fertilizers.[5] Between 1927 and 1932, studies found that organic manures and composts improved plant production and increased their nutritive value and that seeds produced organically had a higher percentage germination than seeds produced with artificial fertilizers.[6] In 1939, twice as much vitamin B1 was found in barley grown organically as in barley grown with artificials[7] and plants grown in sand with artificials contained less thiamine than plants grown in a humus substrate.[8] Plants irrigated with sewer water contained more vitamin B1 than plants irrigated with artificial fertilizers in solution.[9] More recent research into the ecology and biochemistry of soils has shown that soil populations play a major role in making nutrients available to plants, even if these nutrients have been supplied as artificial, soluble, fertilizers that, according to the older theories, should have been available immediately.

It is not too much of an exaggeration to say that the soil acts as a digestive system for plants and in addition to a full range of compounds necessary to plants, a fertile soil also contains many nutrients essential to animal health. In most cases, these substances are found in greater abundance in soils that have been treated organically than in those treated chemically.

An increase in auxins, which stimulate cell reproduction and thus plant growth, has been found in organically treated soils,[10] suggesting that such soils may be inherently more fertile. The level of humus content has been found to be related to the content of vitamins and microorganisms.[11] There are more biotic substances, including vitamins, in the soil adjoining the root system than there are outside the rhizosphere.[12] Vitamins in soils decrease with depth;[13] there is more riboflavin in fertile soils and in soils treated with organic matter.[14]

Studies with animals have shown that stock fed with organically grown feeds produce better with lower rates of disease, sickness and death. Extensive studies have connected animal health with soil condition,[15] but that although they prefer to eat such food, animals can be tricked into eating a nonorganic alternative if sugar is added to it.

Are you still happy with your strawberries?

chapter III

HOW BIG IS THE BARREL?

Can we take from the barrel more than we put into it? For all of us, farmers and consumers alike, this is the question more and more people are asking as they begin to realize that this is precisely what much modern farming attempts to do.

If we think of food as a source of the energy we need to power our bodies, then it is easier to think of agriculture and food distribution as a big barrel. Energy goes in, from sunlight, from nutrients dissolved in water and provided by nature and brought to the plants we grow by the earth's climatic and nutrient cycles— which are also powered by the sun. To this we may add the physical work performed by men and animals, the energy being derived from the food they eat, which in turn is derived from the sun. By far the largest proportion of the energy we use comes from the sun which is, for our purposes, inexhaustible. It is true that one day it will exhaust its own fuel reserves, but the latest estimate gives us better than 750 million years before that happens. If we find ways of adding more energy to the agricultural system, we may expect its productivity to increase. If, then, we come to depend for our survival on this increased productivity, we should consider whether our source of added energy is as inexhaustible as the sun. If it is not, then one day there will be shortages. Energy prices will rise and food prices will rise with them. Then there will be a significant drop in production.

Does this sound familiar? It should, for crops have been destroyed on farms in the United States for lack of gasoline to transport them to the cities. It is not only food production that is affected by energy shortages, but the entire distributive machinery as well.

Of course, there are losses each time energy is used. It is not possible to achieve 100 per cent efficiency—the laws of thermodynamics forbid it. Plants utilize only a few per cent of the energy that theoretically is available to them and according to an ecological rule of thumb, each time food is processed through a consumer, some 90 per cent of its energy value is lost. Thus, for example, animals reared for meat yield only about 10 per cent of the energy they consumed as food. So long as we depend on solar energy, this loss is of no importance, for solar energy is plentiful and it is supplied free.

In the Western world and in Japan, however, a can of soup or a half dozen cucumbers represent many, many times more energy than they will yield in food value. They represent, too, a considerable amount of energy and materials consumed in their production that cannot be replaced—they are exhaustible. It has been calculated for the United States that the fossil fuel energy consumed by tractors is about equal to the total energy yield from agriculture.[1]

The glories of the agricultural revolution and the parallel revolution in food distribution have not been achieved as a result of "break-through" technologies, not even those technologies associated with agricultural chemicals. The real hero is energy, which we squander with a recklessness that is quite staggering. We have learned to tap fossil fuels, which are a form of stored solar energy. With the exception of peat, they were formed during particular geological episodes in the planet's history under conditions which have long since ceased to exist. They are therefore nonrenewable. When we have used them, they are gone forever. We are using them extremely rapidly and our rate of consumption itself increases. Immediately, the limitations that have led to the US energy crisis are caused by the inability of our technology to extract them as rapidly as they are consumed. Soon the limitation will be actual shortage. As prices rise and reserves dwindle, our agricultural glories will fade. We will be forced to acknowledge that we have been living on the world's capital, and treating it as though it were income. It requires no great economic expertise to give oneself the semblance of wealth by spending capital rather than income, but as Mr. Micawber never tired of pointing out, such a course leads eventually, but inevitably, to bankruptcy. We are spending what we should have saved for tomorrow and suddenly we are awakening to the fact that tomorrow has dawned and the day of reckoning has come. Yet still we persist in our attempts to resolve the dilemmas that confront us as though we were experiencing no more than a minor reverse in our fortunes and that our bank balance is infinitely large, as though our problem lies only in carrying the money away from the bank in large enough amounts and that all that is needed to solve it is a larger truck.

All of our economic theories are outmoded, yet we continue to measure farm "efficiency" in terms that take no account of the real world within which farming must function. Can we measure "output per acre," or "output per worker" without allowing for the total energy input and output? How much have we gained in real terms, if we are able to increase the output per man only by providing an energy subsidy equivalent to the labor of several men on the assumption that calorie for calorie fossil fuel energy is cheaper than human energy? The loaf of bread you bought today, produced by modern farming and baking methods, costs as much in energy as you would consume if you were to drive your car forty or fifty miles. Will you derive that much energy from the bread? Of course you will not. As we mentioned in the last chapter, Chinese wet rice farming is estimated to be 6,000 times more efficient than rice farming in the United States.[1] This is an extreme case, but in a similar British study, which has been criticized for the conservatism of its figures which allowed nothing for the transportation of food or of imported raw materials, it was concluded that "a given expenditure of energy on a food producing process achieves a better result if applied to a low intensity system than a high intensity one. Intuitively most people support such a result; a thermodynamicist would expect it."[2] In other words, extensive agriculture is more efficient in its use of energy than intensive agriculture.

The situation was summarized very succinctly by Michael Perelman, an agricultural economist at Chico State University, California, in his testimony before the US Senate Subcommittee on Migratory Labor:[3]

"Let's take one example, energy. Americans use about 18 billion (10^9) horsepower of energy. One man, Fred Cottrell, compared a Japanese farm and an American rice farm. His book was written in 1956, and the studies he used were done somewhat earlier, so what I am going to tell you is even more extreme today than it was when Cottrell wrote it, in spite of the fact that the Japanese are industrializing their agriculture. In Japan, one acre was harvested with only 90 man-days of work, equivalent to 90 horsepower of work. Then Cottrell looked at the studies of an Arkansas rice farm. It took more than 1,000 horsepower just to run a tractor and truck. Consumption of electrical energy on the Arkansas rice farm exceeded 600 horsepower-hours. He did not even bother to ask about the energy used to produce the tractor and truck and other capital equipment.

"We are running into a problem with energy. The whole Alaska find, which is so highly touted according to the oil and gas journals, will probably provide us with enough petroleum to supply our needs for less than 1,000 days. That is less than three years.

"Farms, so-called efficient, large-scale farms, consume more than one calorie of fossil fuel for each calorie of food they produce. That is not an efficient operation."

It has been calculated that a one ton tractor, depreciated over five years, has an annual energy content of 3.4×10^6 kilocalories (Kcal).[4] Even irrigation represents an energy subsidy, and in arid areas it can be considerable. Israel, in one extreme instance, used an energy subsidy for irrigation alone of 50×10^6 Kcal/ha./year.[5]

We are not arguing here that we should return to a system of farming that uses only human and draught animal energy, although it is likely that this would increase efficiency. What we do say is that better ways must be found to utilize the energy available to us. How much must we spend, calorie for calorie, to provide our food? No one really knows, because until quite recently no one thought of looking at the economics of agriculture in terms of energy, rather than cash flow. For this reason, much of the agricultural research that aims at increasing productivity is irrelevant to our true needs. Ways are sought of increasing rates of photosynthesis in crop plants. This can be done by careful breeding—genetic manipulation—to make plants more responsive to inputs of fertilizer and water. This is the basis of the "Green Revolution." But the concentration of nutrients to which the plants respond can be supplied only by artificial fertilizers, which are, in effect, a fossil fuel subsidy, quite apart from the damage they do to the soil and to the nutritive value of the crop. The additional energy that would be required to increase yields in this way is astronomical and the increase could not be sustained, for biological and ecological reasons. Increasing the size of intensive livestock units or the density of crop plants would call for similar increases in the energy added by man. We need to discover ways of sustaining present yields while reducing the energy input. It is our contention that organic farming offers such an opportunity.

The *Ecologist* magazine has suggested, only half-flippantly, that ecologists, who are very concerned with energy budgets, should be renamed economists and that economists, who are concerned mainly with cash flows, should be called "macro-accountants." True ecologists are concerned with the whole of the relationship between man and his societies and the planet.[6]

So Western agriculture, like most other sectors of our national economies, has not considered the possibility of an end to our supply of plentiful, cheap energy. Thus it does not regard its use of energy as wasteful. The aim has not been to produce ever-increasing numbers of calories, or grams of protein, or units of vitamins or other nutrients, at least not specifically. Always the aim has been to maximize financial gain in the naive belief that increasing profits means increasing efficiency. We have done one set of sums but not the other, and the problems we face today are largely the result of our ignorance of this whole dimension to the overall operation. In our eagerness to achieve higher profits, we have ignored the efficiencies of many of the components of agriculture, such as human beings, human communities, the health and safety of workers and consumers. Standards have fallen, but this did not appear in the costings, because the energy input was being increased all the time and the effect was masked. Game plans have been written and rewritten and revised many, many times over the last forty years in order to create a picture of satisfactory profits for investment. The investment was not necessarily the farmer's and the aim was not necessarily to produce food, but to attract capital into big farming and supplying, into labor- (but not energy-) saving machinery and agricultural chemicals. Of course, it has succeeded, although not always in the ways or for the reasons that were envisaged. In Britain, the return on capital invested in agriculture today is related more closely to the rising cost of land than to food prices.

We are told that modern agriculture has "released farm workers for a more productive job" while "increasing the output per farm worker." This is grossly misleading. Modern agriculture has injected huge amounts of energy supported by an energy-consuming industry employing armies of factory workers. This has *displaced* farm workers. They have given up tending horses and gone to work on production lines in the factories that produce the tractors and the chemicals. It is not a simple matter to calculate the true number of workers employed in agriculture at the present time if the production and marketing of "labor-saving" aids to agriculture are taken into account, but a small survey in Somerset in England has shown what common sense would suggest, that the number employed today is not very different from the number employed a century ago. It is just that for most of them living and working conditions have deteriorated.[7] Somewhere between 10 and 11 billion working hours were spent in the United States in 1954, just producing farm goods. By 1967, the industries manufacturing farm supplies in the US were employing 6 million workers. In Britain, there are now more tractors than human beings on the farms. In 1969, with the agricultural work force still shrinking, there were 292,035 farm workers in England and Wales, and 353,690 tractors.[8]

As farm labor forces decrease in size, the number of family farms dwindles, too. The USDA estimates that by 1985 the United States will have fewer than 1 million farmers. If they are correct,

5 ½ million family-owned farms will have disappeared by 1985, and still more farm workers will have been dumped into over-crowded urban areas. This, too, is a global problem, particularly acute in those nonindustrial countries which have attempted to "modernize" their farming along Western lines. The additional and massive human and economic *inefficiencies* this creates should be measured against the increases in agricultural output per worker. It is very evident that in areas where the "Green Revolution" succeeds, the population-food problem is merely exchanged for a population-employment problem which develops rapidly into a problem of rapid and uncontrolled urbanization with no manufacturing industry to take up workers and, consequently, poverty, disease and social chaos.[9]

It is a cost no one can measure, perhaps fortunately. Yet that loaf of bread you bought carries an additional cost in the currency of welfare programs, housing, job training and retraining, and more police and hospitals as human beings break under the strain. During the 1960s, an annual average of 800,000 people were forced to migrate from the rural areas of the United States to the towns and cities.[10] It is a bloodless (more or less) collectivization. As they left, the small businesses that had served agricultural communities closed or went bankrupt and the stream was joined by failed businessmen, unemployed teachers, postmen, bus drivers and others, all seeking work in the towns. As the farms grow in size, so do the towns. Metropolis becomes megalopolis. New York merges with Washington, Glasgow with Edinburgh, London with Birmingham. It was Lewis Mumford who said that the end of this process is reached when megalopolis becomes "necropolis," the city of the dead in which none of the benefits of city life are available to any of the citizens.

Anyone even remotely aware of the problems of densely populated urban areas must know that it makes no sense to continue to increase their size.[11] The migration into the cities from rural areas can contribute nothing to the efficient or economic operation of the urban areas. The cities are suffering from pollution and congestion, taxes are rising, crime rates are increasing and city centers decay as people become affluent enough to move out to suburbs that sprawl farther and farther into the countryside. It may be that at their peak the cities of Western civilization were among the most highly efficient man has ever known. Certainly they were very creative, and the achievements of Ancient Greece and Renaissance Italy were the products of city-states. The size of the cities was restricted and there are still a few towns left whose size has not increased significantly, so that it is possible to see what city life could and should be. Alas, they are few.

It is not only the cities that suffer, of course. There have been studies that have shown that rural communities serving mainly small, family-owned farms were superior to those serving large agricultural units. People had their own ways of measuring efficiency before the advent of cheap energy and their evaluations took account of the quality of the life they led. In 1947, Walter Goldschmidt completed and published a study of two farm communities in California's Central Valley.[12] One community served small farms, the other was in an area dominated by a smaller number of large farms. The community serving the small farms had two newspapers (the other had one), more parks, more independent businessmen, more stores, more retail trade, and twice as many organizations for social recreation and civic improvement; the sidewalks were better, as were the streets and other public facilities, and the standard of living, certainly when measured in terms of the quality of life, was much higher. How do you measure the quality of life? Surely it is possible to see and feel the degree of happiness, contentment and satisfaction within a community? Goldschmidt studied two communities in a nation of thousands.

It is worth emphasizing that farms did not merge into larger units in order to produce more food, in the United States or anywhere else. In the industrial countries of Europe and America, the problem has been over-, rather than under-production. This is true even in Britain, the world's largest food importer. Britain produces less than half of the food its people eat and that half is subsidized very heavily with imports of energy, in the form of artificial fertilizers and feedingstuffs for livestock and fuels to manufacture and power the machines. Yet traditionally a section of the domestic market had been kept open for overseas suppliers who, so it was felt, might retaliate by refusing, or being unable, to buy British industrial goods should the markets be closed to them as a result of increasing home production. Today the British have still not realized the precariousness of their position. As the gap between rich and poor nations continues to widen, markets for industrial products shrink and as, little by little, the nonindustrial countries develop the capacity to manufacture their own goods, markets shrink still further. At the same time world food prices are reacting to shortages and, just around the corner, there is a near certainty of acute energy shortages. British consumers, like their American counterparts, have complained about rising meat prices, but they have not been told that this is a trend rather than an isolated incident. They have not related the "cod war" with Iceland to the need of a small country to assume control over its reserves of a scarce resource, nor considered the implications of this. British military might may conquer Iceland, just as the United States may conquer the Middle Eastern oil states, but a problem postponed is not a problem solved. The British have been told that the era of "cheap food" (it did not seem cheap to them) is over, but only in relation to the revision of agricultural support policies as the country adjusts to the Common Agricultural Policy of the EEC. They have been told little about world demand and supply. On the whole, the leading British agriculturists still believe that because in the past agricultural productivity has been held back, there is infinite room to expand, even to the point where Britain achieves self-sufficiency. There are others who conjecture that one of the underlying reasons for Britain's entry into the EEC is to secure the nation's food base. Were British farmers to be denied the increased imports on which their national expansion is based, it is doubtful whether any significant increase would be possible. The British devoted several centuries to extending their farm area over much of the world and the wealthier of them lived extremely well. Now that the rural hinterland is going its own way, entry into a self-sufficient community is more urgent than many will admit. Even there they are not safe, for Europe, too, is crowded and its farms depend heavily on energy imports.

Small farm units produce more food than large ones. This is a matter of fact. The more labor-intensive the agriculture, the more productive it is likely to be. The Food and Agriculture Organization of the United Nations (FAO) has produced tables listing countries in terms of agricultural output per acre and output per man employed in agriculture and the result is interesting. The "top ten" in terms of output per acre are: Taiwan, the UAR (Egypt), the Netherlands, Belgium, Japan, Denmark, West Germany, the Republic of Korea, Ceylon and Norway. The most productive per man are, in descending order, New Zealand, Australia, the United States, Canada, Belgium, the United Kingdom, Denmark, the Netherlands, Israel and Argentina.[13]

The USDA was established to "help a system of nontechnical, nonmechanized (and for the most part nonchemical) small family farmers cope with an overproduction producing 'unbearable surpluses.'" Large units were introduced and encouraged because there was money to be made, markets which could be developed for industries manufacturing farm supplies, markets for the by-products from the oil and chemical industries, and because of political pressures, a mixture of good intentions and a desire to manipulate a large society. Significantly, public statements from USDA officials describing fewer than one million farmers as enough "to produce our domestic food requirements and supply exports" can be found as early as 1966,[14] which raises the question: is the new USDA estimate that there will be fewer than one million farmers by 1985 an estimate, or has the USDA almost solved the so-called "farm problem" by substituting for farmers the collection of bankers, industrialists, and speculators who are now installing huge, computerized, integrated systems of super-producers, processors and marketers?

It looks as though the European "farm problem" may be solved in the same way. The Mansholt Plan, proposed by the Dutch agricultural economist Sicco L. Mansholt, makes no pretense of helping the small farmer to flourish within a united Europe. "Our aim is to stop small farming—something that cannot be solved by pricing. We must regulate the market by low prices, deficiency payments and structural reform." In addition to manipulated

"force-out pricing and manipulation," the Mansholt Plan pro-
poses, since the US acreage set-aside scheme is not feasible for
Europe, to pay farmers between the ages of 55 and 65 a pension
if they will retire and sell or rent their land to farmers "benefiting"
from the Common Agricultural Policy (CAP). Farmers willing to
earn a nonfarm income equal to their farm income over the CAP
six-year development program will also qualify for "aid." "There
are too many small farmers in the Community, and one cannot
use a set-aside program with small farmers. It is a political
impossibility. First we must reduce the number of farmers. The
structural reform will reduce the land acreage by 6 per cent or
about 12.4 million acres."[15]

Since Europe's small farmers produce abundant supplies of
food—they have produced "butter mountains" and, incredibly,
a "wine mountain"—and obviously the Mansholt Plan seeks to
reduce this output, is it naive to ask why replace small farmers
with big ones? "Our objective is to give social status in accordance
with other industries.[15]" Yet in traditional societies, the peasant
farmer had a status far higher than that of the factory worker.
Indeed, many would-be political reformers have complained of
the peasants' sense of their own importance. Stalin had to kill
thousands. What Mansholt is saying in fact is that he would like
a new game plan and some new faces. He might be saying that
he is just bored with small farmers and finds peasants narrow-
minded—which they are!

Nature always has the last word. As large operators extend
their empires and push toward ever higher yields, they meet
nature head on. As their farms grow larger they need bigger and
heavier machines. These compact the soil, aeration is lost, the soil
structure deteriorates, water penetrates only slowly or even, in
extreme cases of "hard pan," drains horizontally instead of vertically, and yields decline. Nature permits the large-scale farmer no
more time to plant, to till, to harvest than it allows the small
farmer, but when the labor force is smaller in proportion to the
acreage, more equipment is required and ways must be found to
plant earlier and harvest later. The farmer must cut corners.
Usually, this means that heavy equipment must be moved over

the land when it is still wet, which increases compaction immeasurably. The traditional cereal grower allows the weeds in his fields to germinate and then destroys them by cultivating before he plants his seed. If farmers plant earlier, the weeds and the grain seeds germinate together and a weed problem is created. Herbicides must be used to control the weeds. Frequently, American farmers will till and plant a portion of their acres late in the year to buy time for spring work. The land is left exposed to the harsh winter weather, to the winds, snows and rains. Almost as frequently, erosion destroys topsoil which takes thousands of years to replace naturally. So far as this and the next several generations are concerned, irreplaceable humus, plant nutrients and essential soil microorganisms have been lost.

According to the US Soil Conservation Service, "erosion by wind and water removes 21 times as much plant food from the soil as is removed in the crops sold off the land." Every year, the US alone loses one ton of topsoil for every man, woman and child on this planet![16] The quantities are astounding. According to the US Geological Survey, sediments at the rate of 1.3 million tons per day are added to the rivers of the United States. The Mississippi River alone discharges about one-quarter of a million tons of sediment every year into the Gulf of Mexico—almost as much as all the other US rivers combined.[17]

After 49 years of "enlightened" farming, the latest US Soil Conservation Service survey reports that nearly every US farm acre is in need of conservation. US agriculture is the prime polluter of water as run-off fills rivers, channels, reservoirs, dams and streams. In 1971, the US Government spent more than 250 million dollars digging out channels and reservoirs, and 2,000 irrigation dams are now estimated to be "useless impoundments of silt, sand and gravel."[16]

Britain, with its comparatively gentle climate, has escaped the worst of the damage its farming systems might inflict. But it has not escaped scot-free. The removal of hedgerows at the rate of between 5,000 and 10,000 miles a year for more than a decade has left vast areas without protection and in some areas wind erosion is creating problems. Soil "blows" are common. All that is required is bare earth, a few weeks of dry weather in January

Agribusiness concerns can no longer declare that American agricultural practices are more efficient than hand cultivation. Whereas hand cultivation produces many times more energy than it consumes, farming as it is in this country is an energy sink—it uses up more energy than it produces. According to Michael Perelman, the energy value of food crops we consume in the U.S. is about equal to the energy we burn in tractors alone. And much more energy is expended in producing fertilizers, pesticides, and farm equipment, not to mention the manpower needed to run this machinery.

and February and the winds of March will carry away topsoil. Snow ploughs have been used to clear soil from roads in parts of East Anglia and the East Midlands and in the year of the "Big Blow," 1968, winds of 20 knots carried away soil, seed and fertilizer, blocking the dykes on which the area depends for drainage, and costing the farmers up to £25 an acre.[18] Nor has Britain escaped the perils of compaction and the loss of humus. The Agricultural Advisory Council, an official body, said in its report *Modern Farming and the Soil*, published in January, 1970, that "it is by no means obvious to those not scientifically trained what is or is not an unstable soil but it is essential when considering the importance of organic matter that this knowledge should be acquired by farmers. Some soils are now suffering from dangerously low levels and cannot be expected to sustain the farming systems which have been imposed on them."[19]

Large agricultural units require high initial investments in land and equipment. Combine harvesters, for example, can cost as much as $50,000 and yet it may be necessary to use several, working together, in order to complete harvesting in time. With such huge capital outlays large farmers have little choice but to take short cuts at the expense of good soil management, safe, clean, insect control, and high quality food. Livestock farmers are no better off and the pressure on them to exceed the carrying capacity of their land may be irresistible. It is a joke among some farmers that it is the bank that makes their decisions for them, but in truth it is no joking matter. Large farms are often run by cost accountants whose final decisions rest on how well various alternative methods and materials will look in the annual ledger. Inevitably, costs which cannot be related directly to capital investments or to specific gains in income will seldom be approved and so natural processes, which rarely fit neatly into the profit and loss columns, tend to be ignored until they can be ignored no longer.

Vast fields in the southern states of the United States, where lush crops once grew in fertile soils high in humus and organic matter, with active and balanced populations of beneficial microorganisms, now lie fallow or produce poor yields as impoverished farmers struggle with collapsed soils, low in humus and fertility and underlaid with hard pans created by poor management and overdoses of artificial fertilizers. In the Midwest, land which once produced maize so rich in protein that farmers profited by finishing cattle and hogs for market on it, now produces maize that must be supplemented with fishmeal or other protein foods in order to produce adequate weight gains. The appearance throughout the United States of important trace mineral deficiencies in animal feeds and in human food coincides with the increasing use of so-called "commercial fertilizers." Trace minerals are essential to health, but "efficient" farmers have mined out the humus and, with it, essential organic compounds and trace elements at the same time as they have disrupted colonies of organisms necessary for the synthesis and uptake of plant foods, including trace minerals.

Producers in California's famous Imperial Valley, famed for its large farm units and intensive production—and for its massive applications of chemically processed fertilizers and pesticides—face the prospect that their valley may be dying. The salt content of the soil continues to increase alarmingly and in some cases supervisors of major farm corporations will admit frankly that they do not know why yields are declining or failing altogether in certain fields. "Fifteen years ago we went through this Valley and bought what we thought was the best land available. Now good portions of this same land fail to produce a decent crop." "This soil is so loaded with mineral fertilizers that everything is locked up." "Best land in the Valley now is land adjoining a river where you can get good drainage."

Farther north, along California's coastline, prime land and a prime climate that once produced high quality vegetables now produce "market grade" produce in soils where adequate humus content has been mined down to dangerous levels. Nevertheless, few major producers in the area are willing to set aside 60 days out of the 365 to replenish some of the organic matter. Soil samples in the area reveal thousands of acres with high salt content and high pesticide residues. The same samples also indicate serious deficiencies in micronutrients and organic compounds. The soils in San Joaquin Valley are compacted from one end to the other. Declining yields have become a constant complaint, along with the problems of ensuring that sufficient water penetrates the top few inches of soil to reach the plant roots. "Fertility"—or the loss of it—is a general problem in California, and as fertility falls, infestations of destructive soil organisms—pathogens and nematodes—increase throughout the soils of the state, in spite of frequent fumigations and applications of a range of pesticides. So, although some 37,000 very large-scale farmers produced nearly 5½ billion dollars worth of agricultural products in California in 1972, there is some doubt about how long they can continue with the same practices and about the kind of legacy they are preparing for their children to inherit. It seems hardly necessary to point out that deficient, sick soils produce deficient food. Is it farfetched to relate this to the health of consumers? Out of a total of 6,000 children who took part in a nutritional survey in California, 4,000 were found to be suffering from a folic acid deficiency. This was attributed to the high consumption of denatured "convenience" foods and while this doubtless played a major causative role, reason suggests that at least some of the children whose diet was deficient must have been living in households where the dietary regime was apparently sound, containing meat and fresh fruit and green vegetables. Why, then, was their diet deficient?

A generation ago, and for many generations before that, farmers were brought up to "farm for their sons" and to "live as though you may die tomorrow, but farm as though you will live forever." Today it is recognized by leading agricultural chemists that half of the nitrogen fertilizer applied leaches directly into the nearest water and never reaches the plant.[20] Yields of cereals and most vegetables rose throughout the 1950s and peaked early in the 1960s.[21] Present yields are sustained through increasing fertilizer applications, for the time being. The farmers' sons left for the city a long time ago.

It would be a mistake to imagine that large chunks of farmland in the United States are being held by "producers," who care about good husbandry and how well the land will produce in years to come. Tenneco, for example, has made the point bluntly that its land (more than a million acres in California and Arizona) is simply an "inventory," with agriculture "paying the taxes" until inflation makes it profitable for them to sell it. Land is a good investment and its rapid increase in value makes it attractive to speculators. During the last few months of 1972 and early 1973, the "value" of British farmland doubled to an average of about £600 an acre, with some selling at £1,000 an acre or even more. Large conglomerates combine with a host of speculators— banks, doctors (especially doctors, for some reason in the United States—as insurance against the economic ravages of "Medicare" or socialized medicine?), movie actors and other absentee land-owners as a hedge against inflation and taxation. In protecting their own interests, they squeeze out genuine farmers. If they are large enough to establish their own management programs they do so and it is not impossible, although it is unlikely, that their land will be farmed well, as are the estates owned by English Farms Ltd. in Hampshire, England, on behalf of Ronald Lyons, the 44-year-old real estate millionaire. If they are smaller, they work through hired tenant farmers or they engage a management service. Their cardinal rule has been to "produce maximum profit" and they have created a situation where few farm management services or experienced tenant farmers will recommend materials or methods which cannot be related immediately to annual production with the minimum of risk to capital. Indeed, tenant farmers have little choice. Their rent is high and usually they are expected to find half the capital invested, except in the land itself. A bank will assist them with the other half, but the interest will be high and when added to the rent they can afford to take no risks. It is not difficult to find farm managers, tenants or management service supervisors who would prefer to farm organically, were the land their own, but who would never recommend organic methods to a client or landlord. It is more than likely that more insecticides have been used because the possibility of risk existed than because known and identified risks were actually present.

We live in a magical age. We believe our scientists and tech-nologists can solve any problem and we greet each innovation as though it were a wand given to us to increase our domination over natural forces we do not understand. Pesticides are a prime example. When they were introduced, at the end of the Second World War, farmers and consumers alike were mesmerized. They believed that the end of all pest and disease problems was in sight and that there could henceforth be no other "sensible" way to control insects. The memory lingers on, even today, long after the hazards of using them have been exposed. We are left with agriculture's best-kept open secret. For the most part, pesticides have failed where they have been used in large enough amounts or for long enough. Continue to use them and eventually they will fail everywhere. In many cases they have promoted apparently innocuous insects to the status of pest and provided farmers with a problem far worse than that which they set out to solve in the first place. Approximately half of all the pesticides used in the United States have been applied on cotton fields, where the pink boll weevil and the bug have fought the chemicalized armies to a standstill and have won. In Mexico, modern pesticides succeeded in wrecking an 85 million dollar annual cotton production and seriously poisoned hundreds of people in the bargain.[22] California now faces the constant threat of mosquito infestations bearing human and equine encephalitis and malaria, while the cost of controlling insects on the farm with pesticides ranges from $85 to $120 per acre, with the certain knowledge for the farmer that next year it will cost him more and the pesticides will be more necessary, but less effective, than ever. There are still company scientists who will claim that the continued use of pesticides is a necessary evil, but they are less convincing each year.

The introduction and widespread use of high-yielding hybrid seeds in fruit trees, vegetables and grains has transformed millions of acres of hardy, resistant crop strains into genetic weaklings that cannot survive without intensive and wasteful care and protection, requiring expensive, destructive practices and compromises we might prefer not to have to make.[23] This is one of the "magical" technologies we are busy exporting to the nonindustrial countries. Marvin Harris, of Columbia University, a frequent contributor to *Natural History* magazine, is a noted critic of the "Green Revolution." He wrote:[24] "The main problem with the miracle seeds is that they are engineered to outperform native varieties only under the most favorable ecological conditions and with the aid of enormous amounts of industrial fertilizers, pesticides, insecticides, fungicides, irrigation, and other technical inputs. Without such inputs, the high-yielding varities perform no better— and sometimes worse—than the native varieties of rice and wheat, especially under adverse soil and weather conditions.

"Even when the technical inputs are applied in sufficient quantities, certain ecological problems arise, which seem not to have been given adequate consideration before the seeds were 'pushed' out onto the vast acreage they now occupy. Conversion to high-yield varieties creates novel opportunities for plant pathogens, pests and insects. The varieties also place unprecedented stress upon water resources. In the Philippines, for example, tungro rice virus, which was never a serious problem in the past, reached epidemic levels in 1970 and 1971, when more than half of all the rice acreage had been planted with high-yield varieties. I suspect that the modification of natural drainage patterns to provide extra water for paddies with high-yield plants contributed to the severe flooding that accompanied the typhoons in the summer of 1972."

He went on, "You have to be brutally frank with some experts; you have to push them into realizing it: the Green Revolution is a hoax." What he means, among other things, is that the "Green Revolution" has been used as a label to sell agribusiness in the nonindustrial countries without alarming the folks back home over the fact that the main effect has been to force millions of small farmers to sell their plots of land and to replace them with businessmen who will be heavily dependent on industrial products and more closely integrated into the economy of the industrial nations. Like many other forms of apparently altruistic economic aid, the "modernization" of traditional agriculture is extremely good business for the rich. This might not matter were it not for the fact that it adds measurably to the misery of the poor. In any case, the boom may be nearing its end. As India struggles with the third successive failure of the monsoon, it is said that already the "Green Revolution" has passed its peak in that country[25] and the FAO Director-General, in his introduction to the 1972 edition of *The State of Food and Agriculture*, the FAO Yearbook, refers to the parable of the seven lean years as he explains the continued deterioration in world food supplies.

As Europe bustles about its business of repeating mistakes made in the United States, American citizens, somewhat confused and frustrated by more than forty years of experience at variance with the propaganda, have begun to examine modern agricultural practices in a new perspective. They are questioning the desirability of large units. Lost opportunities, congested communities, high food costs, contaminated environments, polluted waters, extremely high transportation costs per food dollar, lost flavor and maturity and questionable nutrition have posed the question: What is modern agriculture doing for us that was not done better before? People are realizing that today the main qualification for opening a large farm unit is no more than the possession of enough money to take over sufficient land, equipment and stock. From then on, departures from the highest standards of husbandry are rationalized by saying "these compromises are necessary because we are big operators." It is a circular argument. No one has so far explained why big units are better for society, or why a system based on small units would not produce and distribute food just as effectively as big agriculture, and without such an abrasive effect on the ecology, on communities, cultures, people and the quality of food.

The mass handling of food creates problems and costs which are beginning already to erode the primary advantage of volume. Transportation, in most cases, now costs many times more than the food itself. It now costs one dollar per case to ship tomato juice from California to New York in carload quantities. Fresh produce ranges from 4¢ to 12¢ per pound for transportation. A crate of lettuce, which sells for $1.75 to $2.25 for 24 heads, may cost as much as a dollar a crate to transport 300 or 400 miles. People living in rural areas in Britain buy vegetables staler than those sold in the large cities, even though they may have been grown no more than a mile or two from where they live. Almost all fresh produce is handled by a few distribution centers, the largest being London's Covent Garden market. The countrydweller must wait while produce is taken to town and then brought back again, bought by the purchasing branch of the local supermarket chain.

As fuel shortages continue and we learn, because we have no alternative but to learn, to use fuel more thriftily, costs are bound to rise. Fuels, lubricants, industrial products and chemical products will all become more expensive. Therefore food will become more expensive, but the relative increase in costs will depend directly on the farming techniques used. Farming on a large scale, using large numbers of large machines and heavy applications of artificial fertilizers and pesticides, will be seen to be an expensive way to farm—which it is. When to the production costs are added the costs of transportation, storage, processing, packing and handling, the difference in cost between the large-scale production and distribution operation and the small-scale farmer, using fewer machines, no artificial chemical products, and selling locally, will become very evident. The consumer will receive less and less food for every dollar or pound spent.

Until that not-far-distant day when we are forced to come to terms with the real world in which we live, attempts will continue to convince us that large-scale farming is more efficient, and that it means cheaper food. The secret of the conjuring trick is the economic sleight of hand on which our agriculture is based. We expend perhaps a hundred thousand times more energy producing the food and bringing it to you than the food will yield. So long as we are able to believe that this energy is provided almost free of charge, the illusion will hold. It will not hold for much longer.

chapter IV

THE ORGANIC ALTERNATIVE

The organic alternative is not always easy to understand. How could it be? The world is complex and agriculture is engaged in some of its most complex processes. Any attempt to deal with the subject holistically is bound to be difficult, conceptually if not in practice. To many people it suggests, vaguely, that food is "produced without chemicals." Indeed, organically grown food is often marketed with a label that says no more than "compost grown," "pesticide free," or "grown without chemical fertilizers." It is true, of course, that organically grown food is grown without the use of agricultural chemicals, but it is also true that all living processes are organic, by definition, and that if we are to adopt a reductionist rather than a holistic approach, all plants are the result of chemical reactions and that the concept of growing food without chemicals is meaningless. Organic chemistry and biochemistry are the sciences of the chemistry of living processes. To a chemist, everything is explicable in chemical terms, at least so far as food production is concerned. If "chemical" farming is inadequate, this is due to deficiencies in our understanding of chemistry. It does not challenge the chemical interpretation of nature. Should we wish to challenge this interpretation, and it is the view of those who support the organic alternative that it should be challenged, then we must cast our net wider.

In fact, there is a great deal more to it than that, and to some extent a preoccupation with "chemicals" is a red herring. The aim of organic farming is to produce food using methods and materials that make the most efficient use of nature's resources and in ways that are compatible, over a long period, with the laws and conditions of life on this planet. This view of life and of food production was, perhaps, expressed best by Sir Albert Howard, who wrote: "Instead of breaking up the subject into fragments and studying agriculture in piecemeal fashion by the analytical methods of science, appropriate only to the discovery of new facts, we must adopt a synthetic approach and look at the wheel of life as one great subject and not as if it were a patchwork of unrelated things."[1]

This is an attack on reductionism and a plea for holism, but, commendable though it is, it is far from easy to reduce to a neat definition. Indeed, it defies reduction! Yet the need for a definition exists for certain, limited purposes, such as consumer protection, and in May, 1972, the United States House of Representatives was presented with a bill to amend the Federal Food, Drug and Cosmetic Act to regulate the advertising and distribution of organically grown and processed foods. The bill contained the following definition:[2]

"For the purposes of this section:

"(1) The term 'organically grown food' means food which has not been subjected to pesticides or artificial fertilizers and which has been grown in soil whose humus content is increased by the addition of organic matter.

"(2) The term 'organically processed food' means organically grown food which in its processing has not been treated with preservatives, hormones, antibiotics, or synthetic additives of any kind.

"(3) The term 'organically growing or processing food' means growing or processing food for distribution in commerce as an organically grown or processed food."

The organic alternative is not one, but several related alternatives, and their implications. The pattern of relationships between living things and the inert materials on which they depend is like a vast web. One strand links with another, which leads to several more, and more, and so on. For example, organic farmers use waste organic materials—manures, straw and even garbage —to feed their crops and to build up the humus content, and fertility, of their soils. If we are not altogether happy with words like "chemical" and "organic," we are even less happy with "waste" when this term is applied to the product of one stage in a cycle that should lead it in as the raw material of a later stage.

Before plant nutrients present in organic matter, such as mineral and humic compounds, can be recycled from the raw material, there must be a breaking down, a decomposition. If organic matter is applied or worked into a soil in its raw state, decomposition will take place eventually. It is a natural process brought about by soil animals, such as earthworms and certain beetles and spiders, by microbes and by oxidation. Some farmers use their material in this way with satisfactory results, but for others it may take too long, it may introduce unwelcome weed seeds which it encourages to germinate, or, in some cases, it may introduce harmful bacteria. So the organic farmer may wish to use methods that speed up the decomposition of his organic "wastes" before he applies them to the soil. He has several methods to choose from and they are all called composting.

It is this interest in the decomposition stage of the nutrient cycle —or the life cycle, if you prefer—that characterizes organic farmers. It is quite logical. The growth of plants as primary producers to feed a hierarchy of consumers from herbivores through omnivores (such as man) to several levels of carnivore is part of a cycle that also includes their death and reduction to the simpler chemical compounds of which they were constructed. Chemical farming seeks to accelerate the cycle at the growth phase by supplementing the supply of nutrients directly to the plant or to the herbivore. Organic farming seeks to accelerate the cycle at the decomposition phase by improving the rate of breakdown of organic materials. This approach is more subtle, less immediately obvious, but it has clear advantages.

"Simply stated, composting is the biological decomposition of the organic constituents of waste under controlled conditions. . . . It is the application of control that distinguishes composting from the natural rotting, putrefaction, or other decomposition, that takes place in an open dump, sanitary land-fill, in a manure heap, in an open field, etc."[3]

Under controlled conditions, all organic matter can be composted, which is another way of saying that natural processes can be harnessed efficiently and safely to unlock and make available rapidly the plant nutrients, the minerals, acids, and other humic compounds, vitamins, proteins and antibiotics taken up by plants or ingested by animals; at the same time, a large amount of humus is prepared and returned to the soil through composting. As Dr. W.E. Shewell-Cooper, the British authority on organic gardening, is fond of saying, "If it has lived once, it can live again."

Most modern techniques of farm or garden composting are based on Sir Albert Howard's Indore method, named after the research station in India where he developed the process, and not because it is to be made in the house! Animal manures and vegetable matter are stacked in layers, with lime added at intervals to keep up the pH. The heap must have a "critical" mass, which usually means it should be not less than about five feet by five feet by six feet high when built. It will compact down to a height of about four to five feet. It must be moist, but not too wet, and it must be aerated. This can be achieved either by building it off the ground so that air can enter from below, or by driving stakes through it vertically and then removing them to leave holes, or both. Fermentation will begin quickly and the temperature of the heap will rise. When the heap cools, it is turned by removing the material from the top and then rebuilding the heap on top of it, so what was at the bottom is now on top. This should cause the fermentation process to begin again and, once more, the temperature will rise. When it falls the heap should be left for a further period. The initial high-temperature fermentation, which is caused by the activity and proliferation of thermophilic organisms, will be followed by a slower, anaerobic putrefaction, which is also necessary for complete decomposition. The entire process should take no more than a month or two, depending on the season— obviously it is easier to sustain a high temperature during summer than during winter. The resultant compost is dark brown, sweet-smelling, light, friable humus.

On the farm the system may be simplified and, in any case, it is far less onerous than it may sound. Whereas garden compost is made in small heaps, usually contained within boxes, or sometimes wire cages, of one kind or another, farm compost is made in windrows—long stacks, rectangular in cross section, but still roughly six feet deep by five feet high. The building of the windrow is simple with equipment found on most farms. While he was Farm Director for the Soil Association, Douglas Campbell developed what may be the simplest method of all. In early spring, when the cattle left the covered yards for the fields, the yards were "mucked out" using a tractor with a front-loader attachment, onto a conventional muck-spreader, which is no more than a farm trailer with a flailing device at the rear end which throws out the contents of the trailer as they are pushed toward it. The muck spreader made the heap, the flailing action ensuring adequate aeration. The heap was shaped by a three-sided "mold" made of corrugated iron, which looks like a small hut open at both ends. This was towed on skids behind the muck spreader, so that the material was thrown into it. As soon as it was filled, the whole unit of tractor, trailer-spreader and "hut" moved on a few feet. As it worked, a second man (the first was driving the tractor) forked green matter into the trailer to be mixed with the manure, and added lime. The entire operation took three men (the third drove the front-loader) about a day. The compost can be turned, again using the front-loader, but the Soil Association heaps were left to mature without turning. Some organic farmers, however, do turn the heaps. The compost made by this method is rougher than garden compost, but perfectly satisfactory for farm use.

It is often said that one man's need is another man's opportunity. All farmers need organic matter, although some are more aware of this need than others. Organic farmers know that the nutrient and humus content of soils may be seriously depleted by years of intensive reliance on NPK. For this reason, organic farmers and those who advocate the organic alternative, propose composting, on an industrial scale, as the obvious alternative to the massive, costly and in some cases eventually insoluble problems of disposing of billions of tons of organic wastes every year. Garbage, animal manures from intensive units and feed lots, human sewage, all tend to be concentrated in relatively small areas and unless they can be recycled they must be burned or dumped—on tips, in holes in the ground, in rivers, in the seas and in lakes. It is not at all unusual to hear organic farmers discussing problems of urban solid waste disposal and talking of the relative costs and benefits of composting, as opposed to land-fills, and talking knowledgeably about water pollution. The Soil Association has amassed more information on the composting of town wastes than any other organization in Britain. In the United States it is Rodale Press, pioneers of organic farming and gardening, which sponsors conferences on the subject, which farmers attend, and which publishes the most authoritative journal on the subject in the English language and possibly in any other, *Compost Science*. What most people regard as an embarrassing waste, to be disposed of as conveniently as possible, they regard as a raw material that is being squandered senselessly and that damages the environment in the (bad) bargain.

They are right, for the technology that would enable us to re-cycle this material exists already and is being improved all the time. Methods are many and various, but usually they begin with the sorting of materials to remove those not suitable for treatment. Then they are shredded or milled to reduce their bulk. There are several devices for this purpose, but in general those that work well in Europe are less successful in the United States, owing to the higher proportion of bulky material in US garbage. Then they are composted, with or without the addition of partly dried sew-age sludge. The aim is to achieve a high-temperature fermentation to accelerate the natural process and also to kill pathogens quickly. Depending on the space available and the capital avail-able for investment in the process, the material may be matured in windrows or it may be subjected to a more sophisticated treat-ment, such as the Dano process, which is the one most widely used in Europe. Here the material is injected into large drums which rotate slowly, so that the material mixes constantly. Very high temperatures are achieved and a fermentation that would take weeks in an outdoor compost heap can be completed in a few days. The compost is then stacked in windrows to mature.

There are objections to the recycling of urban waste. For years municipal composting has been rejected because it was considered uneconomic. Regarded purely as a waste disposal process, this was true while land was available for tipping. After all, nothing is cheaper than emptying everything into a convenient hole in the ground. However, more and more cities are finding tipping sites more expensive to buy, because of rising land prices, and to use, because they are farther from collection points. The most popular alternative is incineration, but as air pollution standards become more restrictive, composting, which causes less pollution of air and water than any alternative disposal process, becomes more attractive. There are objections, too, to the glass which remains in the finished compost in small fragments. This, it is argued, could present a hazard to workers who must handle the material, yet there appears to be no recorded incident of an injury caused by this glass. The reason is that glass itself is biodegradable—it will decompose and the bacteria that attack it work first on the rough edges, smoothing them.[4] Other objections to the possibility of toxic residues, especially of heavy metals, originating in industrial effluent that has been added to sewage, surviving the composting process, are more serious, but not insoluble—at any rate in theory.

Someone not familiar with the implications of the organic alternative might find it surprising to attend a meeting of organic farmers at which so much attention was paid to often highly technical discussions on urban waste disposal. Surely, they should be talking about the nutritive value and flavor of foods?

In a way, that is precisely what they are talking about. The way to improve the quality of foods is to increase the amount of organic matter returned to the soils that grow the foods, and composting is the technique by which this may be done. The "mineral" theory is bankrupt. It suggests that plants require only measured amounts of nitrogen, phosphorus, potassium and calcium, plus a few trace minerals, for healthy growth. More recent research has shown that plants need at least 60 minor minerals, and there may well be more that so far have not been identified, as well as humic compounds. They need, too, an association with various types of soil microorganism. Only when these conditions are fulfilled will the farmer be able to produce highly nutritious, flavorful foods, and these conditions will be fulfilled only in soils with a high humus content. So garbage, manure and sewage and ways to compost them are as relevant to the organic farmer as are metals and ways of mining and refining them to the car manufacturer. They are raw materials and the techniques for processing them. They are also one strand of our web and they link with other strands until we find ourselves considering a complex so large that it affects the daily lifestyle of every one of us. When you throw away garbage, when you use the lavatory, do you think of the effects on the environment? Probably you do. But do you think, also, about ways in which you might make it a little easier for those who will take your "waste" products and turn them back into food?

This broad, holistic approach is far more important than the name of the alternative. Nowadays, some young people call organic farming "ecological" farming. If, by ecology, they mean the relationship between human communities and their total environment, perhaps this name is better. It comes closer to expressing what we mean and what organic farmers themselves have been saying and doing since long before the word ecology became fashionable. Both words suggest, or should suggest, a better life for all of us and most of all for those who must follow us as tenants of the planet.

The organic, or if you prefer, ecological, approach to the solving of everyday problems is as fundamental as the words themselves suggest. *We must learn to understand and work with natural processes or we will be destroyed by them.* When we begin to apply the organic philosophy we find ourselves asking such questions as: *Does this solution to my problem, or this innovation, increase the stability which the existence of the problem suggests has been lost? Does it offer a permanent, or at least long-term situation in which the problem will not recur? Does it create more problems than it solves?* We must learn to watch out for those solutions that are not solutions at all, but which merely transfer the problem to someone else, or postpone a true solution to people somewhere in the future.

Now look again at some of the "solutions" modern man has found for the problems he faces.

The amount of topsoil that is removed from U. S. land by wind and water erosion is astonishing. The present commercial farming practices—which do little to improve the humus content of the soil, leave much land exposed to harsh winter weather, and compact the land with heavy machinery—are responsible for much of this erosion.

Conservation measures practiced by organic farmers, such as cover cropping, mulching, and humus-building, are needed on almost every agricultural acre in the entire country.

Problem: Our teeth are rotten because of the poor food we eat. Solution: Train more dentists.

Problem: Our cities produce large amounts of garbage and sewage. Solution: Find more holes in the ground and build bigger pipelines to carry the sewage to the sea.

Problem: Our cities are choked with traffic. Solution: Remove large areas of city centers to build bigger roads and parking lots.

Problem: Air pollution in some cities is so serious that it has an immediate adverse effect on the health of citizens. Solution: Wear gas masks.

It is an amusing game for the children to play on a wet Sunday afternoon and it can go on for a long time. Invariably, the problems are caused by modern technology and the solutions, based on still more technology, create more problems than they solve if, indeed, they solve any at all. Should you still need convincing, ask yourself how appropriate are many of our responses to the problems we face, and for how long our solutions will last. London has cleaned its air, for example, by changing from coal to smokeless fuel for domestic heating and by building higher stacks for its factories. True, the air is cleaner, but the production of smokeless fuel causes serious pollution of the air in the areas where it is manufactured, and the higher stacks do not remove pollutants, they simply ensure that they are carried farther before they fall— in this case to Scandinavia. The Thames has been cleaned, too, partly by using boats to carry sewage out into the North Sea, where it is dumped. If we hide our wastes out of sight, how long will it be before we run out of places to hide them and they begin to accumulate on our doorsteps? In some cities, such as Liverpool, the pumping stations that speed our sewage on its way to the sea are below sea level. Were they to fail, the streets would be flooded with raw sewage to a depth of several feet in a matter of hours. If we rely on gas masks to "solve" an air pollution problem, how long will it be before we must sleep in them and how will we eat?

Now ask yourself for how much longer we can continue to base our food production on the mining of nonrenewable mineral and fossil fuel resources, without even attempting to recover the billions of tons of materials we waste?

You will find that your view of the world you live in is changing. You will find yourself asking and trying to answer difficult, profound questions. Where have we come from? Where are we now? Where are we going? What have we done? Do we like what we do? Can we keep it up? Should we change? How? When this happens, you are beginning to think organically.

If you talk to organic farmers, or listen to them talking to one another, if you read the books and periodicals they read, these questions will keep recurring. They underlie everything.[5] Let us take a very simple example. What is the best way to mill flour? As things are today, mills grow larger and larger to handle the grain with the minimum of labor and the maximum of energy. The germ of the wheat is removed because it contains oils which do not keep well and because the germ interferes with the milling machinery. Synthesized nutrients are returned to compensate for those removed for technological convenience, while more substances are added as preservatives, stabilizers, bleaches and so on, all compromises "necessitated" by the process itself. Is this the best way? Can such large mills provide the consumer with as full a range of different flours as the farmers are able to provide the grain for? How large does any operation need to be before its efficiency begins to diminish and it experiences diseconomies of scale? Would we have better, fresher flour, and a wider choice, if we had more, smaller mills, each serving its own locality? In the United States and in some European countries, it is possible to buy really small home mills. Do they provide people with the best solution of all—their own flour, ground fresh to their own specification? Home mills are not yet available in Britain, but even there some people manage to grind their own flour, using the grinding attachment of a food mixer. It works well enough provided the machine is not run too fast. It looks as though the "modern," "efficient" way to provide fresh flour is for the housewife to grind it for herself at home. Thus buying prepared flour becomes outdated and we have come full circle: this is where we began!

Now, you may find this idea amusing, but perhaps this is because you are not fully aware of all the compromises of quality that are made in the course of large-scale milling and distribution of flour. Quite a number of organic propositions have seemed amusing until they were examined in a wider context. Until quite recently, for example, such advocates of organic farming as J.I. Rodale, of Rodale Press, and the Soil Association were ridiculed for suggesting that modern chemical pesticides were disturbing ecological balances, so producing more problems in the long term than they could solve in the short, while at the same time creating an unnecessary hazard to human health. Pesticides still have their enthusiastic supporters, of course, and they still scoff, because they cannot believe that a risk exists until catastrophe provides them with proof they will accept—reluctantly. Their attitude is the antithesis of the organic idea of relating a practice to the entire web of living relationships before applying it on a wide scale in the field, and so before serious damage can be done. The public relations men are shifting their ground as they try to make the best of a bad argument. ICI, one of Britain's largest agrochemical manufacturers, published an editorial in one of its journals called "Agricultural chemicals as ecological tools," in which it said: "Pesticides have always been instruments of ecological change— blunt maybe and sometimes ineffective and even dangerous, but their very survival implies that they have found a role in the regulation of the human ecosystem."[6] Well, that is one way of putting it! In fact, there is little known today about DDT, for example, that could not have been known in 1942. Fortunately, more and more scientists agree, and more and more are proposing, that ecology must become a "theoretical, predictive science."[7]

In their wilder flights of fantasy, the opponents of the organic alternative have accused it of being part of a sinister plot to subvert the food production capability of the entire western world.[8] This may sound absurd, but there is wide acceptance of the view that a switch to organic farming would lead to mass starvation in the United States, Britain and Europe. Our memories are short. Prior to about 1920 to 1930, farmers had few agrochemicals and many of them used none at all. They had fewer machines and those they had were smaller. They had little or no modern irrigation. Yet they provided the economic base to build and feed all the cities and most of the towns and villages of Europe and North America. Even today, more than 82 per cent of US farm land receives no pesticides, and more than two thirds of its arable land receives no agrochemicals at all.[9] Organic farming has many centuries of proven achievement to its credit, rather than forty or so dubious years. We *know* it can form a reasonably stable system while providing more work opportunities for more people.

Two thirds of the world's farmers farm organically because they cannot afford chemicals. The world food problem is not the result of the inefficiency of their farming systems, but of the pressure on food supplies exerted by rapidly rising populations, compounded by high levels of food wastage and social and economic inequalities. Fertilizers will not solve the problem. The largest national population in the world, amounting to something like one human being in every five on the planet, lives in the People's Republic of China. It is fed by what amounts to a system of small, organic farmers, linked together to feed the people in their own locality.

Organic farming lends itself to a system based on small, family units, and its consumption of materials imported to the farm is small. Thus, there is less money to be made from the sale of goods and services to organic farmers. Perhaps this provides a clue to the motivation of its opponents? It is argued, for example, that the agriculture of the United States can be sustained only by the application of billions of pounds of chemical pesticides into the environment and into food each year. The United States consumes (perhaps the word is more apt than the economists intended!) almost one billion pounds of pesticides per year, applied as we have seen to certain crops grown on a limited acreage, but applied in very large doses. It hardly amounts to an issue in itself, except that it is kept alive for the sake of the profits of the chemical companies and because of what must seem to them as their hope against hope, as a market capable of considerable expansion. In the United States it is possible to substitute land to compensate for any loss in production resulting from abandoning pesticides and it has been estimated that "eliminating pesticides would involve crop losses of only 7.1 per cent over losses incurred with pesticide use."[10] Is a loss of less than 10 per cent enough to justify the health hazard, the contamination, the cost, or to off-set the fact that millions of acres have been "set aside," taken out of production, "in order to control surpluses?" The relationship between pesticide use and crop production has been broken down by Professor David Pimentel in the March, 1973, issue of *Environment*. His figures are illuminating (see table).

If these results are studied in conjunction with the findings of other investigations, it becomes evident that increasing the amount of land in production reduces disproportionately the chemicals that are "needed." In the United States, a 1 per cent increase in crop land would reduce the insecticide "requirement" to produce the same total output by 6.5 per cent.[11] Unfortunately, there have been no thorough, impartial studies, and none can be expected for some time, of the comparison of output from soils managed organically, without any use of pesticides, but such unbiased, albeit fragmentary, information as we have indicates that a reduction in the heavy use of agricultural chemicals tends to increase farm incomes and that it "leads directly to a reduction or reversal in the emigration of farm populations to cities."[12]

The principal method of insect control used by organic farmers is a healthy, biologically active, fertile soil and the rather obvious fact that a healthy soil produces healthy insect- and disease-resistant plants and animals. How well this works for a farmer depends on how long he has been farming his land organically and how deliberately he has worked to rebuild his soil. The concept is basic to organic farming and so it has been attacked repeatedly by critics who maintain that there is no evidence to support the "theory," that it is based on "muck and mystery." This is partly true. There is very little "evidence" to show why this effect should occur. On the other hand, it is more than a theory, since it is based on the observations and experiences of many farmers in many countries, working independently of one another. In this way it has been verified many times. If we have little idea of why it should be so, perhaps this is because for the past thirty years few scientists, if any, have measured and examined closely crops grown by organic farmers.

How much evidence do we need? Because we do not know why a phenomenon occurs this does not mean to say the phenomenon is imaginary. We do not understand a great deal about gravity,

US CROP LOSSES DUE TO PESTS, 1904 to 1951–60*

Causes of crop loss	1904	1910-35	1942-51	1951-60	Losses if no pesticides were used
Insects:					
Losses (billions of dollars)	0.4	0.6	1.9	3.8	4.8
% age of crop lost	9.8	10.5	7.1	12.9	16.3
Crop diseases:					
Losses (billions of dollars)	NA	NA	2.8	3.6	4.2
% age of crop lost	NA	NA	10.5	12.2	14.2
Weeds:					
Losses (billions of dollars)	NA	NA	3.7	2.5	3.0
% age of crop lost	NA	NA	13.8	8.5	10.2
Total loss:					
Losses (billions of dollars)	NA	NA	8.4	9.9	12.0
% age of crop lost	NA	NA	31.4	33.6	40.7
Potential production value in billions of dollars	4.1	5.7	26.7	29.5	29.5

*This table originally appeared in "Realities of a Pesticide Ban," David Pimentel, *Environment,* 15(2):28.

or the way it exerts a force of attraction between bodies, but we do not need to jump from the roof of the Empire State Building to prove that, in spite of this, gravity does exist. There is substantial evidence, based on good, solid investigation, which continues to accumulate and which shows that the soil conditions produced by conscientious organic soil management promote those conditions of structure, moisture retention, soil organisms, organic compounds and other nutrients that are necessary for healthy plant life and which produce healthy animals. Investigations of fertile soils reveal the presence and availability of vitamins, particularly those in the B group, amino acids, major and minor minerals, biotic or organic compounds including antibiotics, extremely active and abundant colonies of microorganisms, and a high humus content with plenty of decaying matter.[13] Where the humus content is low, there are smaller amounts of advantageous microorganisms, organic compounds, trace minerals, plant nutrients, vitamins, antibiotics, and the structure and moisture holding capacity is much reduced.[14,15,16,17,18]

Nutrition of plants, animals or humans is a far more complex business than many scientists imagined even quite recently, and it is absurd to suggest that all the substances important to a sound diet have been identified, far less their function defined. This was demonstrated dramatically some years ago by a Czech worker in Prague. He removed all the fats from an experimental rat diet and with them all the fat-soluble vitamins. He then returned the fats in the form of pure fat, together with all the fat-soluble vitamins in their pure form and correct amounts. Within a few generations his animals had died out, sick and infertile. Clearly, there were other substances in the fats of which he knew nothing. The diet of a small rodent is much simpler and much easier to control than that of a plant growing in soil. Numerous studies have documented the ability of various microorganisms found in fertile soils either to synthesize or to transmit vitamins, amino acids, proteins, minerals, organic compounds and water.[19] Other studies have traced the uptake of these substances by plants and have produced substantial evidence to show that under laboratory conditions some plants appear to be unable to synthesize certain substances which they take up from a fertile soil, and that they respond well to humus, certain nutrients, organic compounds, acids and vitamins which had been withheld from a testing solution or in which a soil was deficient.[20,21] There are more recent studies which confirm Sir Albert Howard's observation that if a plant is to produce well and resist disease under normal field conditions, there must be a symbiotic relationship between its root system and certain soil microorganisms.[22]

Even so, it is still difficult to show conclusively that foods produced organically are nutritionally superior to or have better flavor than those grown with chemical fertilizers. This is largely because there have been very few comparative studies of foods grown on similar soils but under different systems of management. If we accept that analytical techniques are unsatisfactory, since we can analyze only for substances we know and can identify, the only valid technique left is the feeding trial. To compare foods they must be fed to animals and the health of the animals compared. The Soil Association prepared the way for a study of this kind by farming land under different regimes for many years. Differences were observed in the performance of livestock, but the farming management was open to a number of criticisms and at the time of writing no final report of its experimental work has been published. However, it did at least provide a base on which others might build and it was never possible for it to finance the very expensive fundamental nutritional research that was needed if its material was to be investigated thoroughly.

To do this it is necessary to acquire a population of animals large enough to make comparisons that are statistically significant. In order that genetic characteristics in the stock that might account for differences in performance can be eliminated, the colony must be bred from foundation stock of known genetic characteristics. Extraneous environmental influences must also be excluded, so that the animal unit must have an environment in which temperature, humidity, lighting, noise and infection are under rigid control. In 1971 the Soil Association's research role was taken over by the Pye Charitable Trust, which has funds adequate to undertake work of this kind. The animal unit has been installed, the colonies of small animals are under observation and the comparisons which are being made are beginning to show what seem to be differences in health that can be traced only to the way in which the animals' food has been grown. So far no results have been published.

Such analytical work as has been done shows that the nutrient content of chemically produced foods varies greatly from one field to the next and even from one part of a field to another.[23] It has been established that superphosphates in the soil can inhibit the uptake of zinc, especially if the soil is low in humus,[24] but that zinc uptake increases even in the presence of high superphosphate or salt concentrations, when humic acid is applied.[25] We know that the soil content of humic acid and other organic compounds increases in proportion to the organic matter content[26] and that plants grown in soils high in humic matter contain more trace minerals.[27] We know that certain fungicides, herbicides and fumigants used to control fungus in orchards and to sterilize soils destroy colonies of essential mycorrhizae. Plants will not grow well in soils deficient in mycorrhizae without heavy applications of chemical nutrients, whereas plants grown in soils with adequate mycorrhizae and humus show the same vigorous growth with much less fertilizer.[28] The soil is a far more complex substance than it seems and, even today, soil ecologists are only at the beginning of their investigations of micropopulations. In a handbook produced as part of the International Biological Program in 1971, one worker pointed out that "the relatively fundamental question of estimating numbers of soil organisms is very far from being answered in many cases, while microbial metabolism in the field has not been measured directly at all."[29]

There are independent studies, some of them far from recent, that show that plants grown with organic fertilizers can be more nutritious.[30] A recent study, an important one for a generation suffering from iron deficiency, demonstrated that potatoes grown with barnyard manure contained twice as much iron as potatoes grown with mineral fertilizers.[31] Private analyses by farmers, publishing companies, private laboratories, and others show repeated instances of foods produced in soils under conscientious organic systems of management being significantly higher in their content of essential nutrients and in the case of protein foods, amino acid patterns are superior. It is known that the fat structure in the meat of animals reared on free range, eating a natural diet, is different from the fat structure of meat from animals reared intensively on a fortified diet, and that this difference is significant to human health.[32]

It is difficult to convince those who do not wish to be convinced and so the matter remains controversial, as is so much of science. There has been little inclination on the part of university departments of agriculture to research in this area, although the German Federal Institute for Research into the Quality of Plant Products (Bundesanstalt für Qualitätsforschung Pflanzlicher Erzeugnisse) at Geisenheim-am-Rhein is investigating nutritional quality under its director, Prof. Dr. Werner Schuphan. Unfortunately, none of his papers has been translated into English since his book was published in 1965,[33] and that described his work only up to about 1956. His German is not easy to read, or to translate. In Britain there is the Pye Research Centre, to which we have referred, and in the United States the only project at present is being undertaken at the West Virginia University. So far this has shown that nutrients vital to human health decrease with continued applications of chemical fertilizers. At the time of writing, this project is in danger of losing its funding.

It is easy to become lost in an important, but highly technical debate about whether or not organic systems of husbandry enhance nutritional quality, but if we turn the question around the evidence is much clearer. How nutritious are the persistent pesticides that accumulate and mix in our bodies? Are they safe? No one knows for certain, but such evidence as there is suggests we would be better off without them. DDT in chickens will inhibit egg production and weight gain in proportion to the amount of DDT present. There is ample evidence of physiological and behavioral disturbance caused by organochlorine residues in birds of prey.[34] Very often, hormonal balances were changed, leading to sterility. In one investigation, the second generation of birds whose parents had been exposed were sterile, although there was no other indication that either they or their parents had suffered. There is a wealth of evidence about the harmful effects of nitrate fertilizers, some of it going back as far as 1929.[35] Phosphate fertilizers can alter the nutritional composition of certain crops under some conditions.

Nitrate fertilizers can actually be poisonous if they lodge in plant tissue in their nitrate form. Some crops, such as spinach, are especially prone to this. Bacterial action in the rumen of animals can bring about the reduction of nitrate to nitrite, which is able to pass through the gut wall. Absorbed into the bloodstream, it forms a stable compound with hemoglobin, methemoglobin. This prevents the formation of the unstable compound oxyhemoglobin by which oxygen is transported to the cells of the body. In extreme cases death can result from asphyxiation. A further byproduct of the reduction of nitrate to nitrite is hydroxylamine, which is also able to enter the bloodstream and which plays a principal role in the destruction of vitamin A, as well as having a hemolytic action which can give rise to anemia. The bacteria responsible for the reduction of nitrate are also found in the gut of human infants, and babies have died from eating food containing excess nitrate as a result of overzealous fertilization.[35,36]

Organic farming favors the small, family unit and it may well be that this is the reason there are so few research programs to examine its techniques and results. It may work in two ways. Most university research is funded from outside and the tendency is for more and more to be paid for by industry. Naturally enough, industry will finance only that work which promises eventually to lead to increased turnover based on rising sales of its products. Small farmers offer a poor market. The second reason strikes deep into our culture. We are all inclined to patronize the family farmer, or the peasant, which is what we often mean. The very word "peasant" is often used in a pejorative sense in English and we are able to use it only with great care. Meanwhile, civil servants and the directors of government research establishments find themselves seduced by the same easy acceptance of the general view that "big" is synonymous with "good" and "bigger" with "better." Hence political programs which aim to increase the scale of any operation are likely to seem more attractive than those which seek to scale them down, or even to preserve the status quo. Policies are devised by civil servants in consultation with the industries likely to be affected. In the case of agriculture, the industry is represented on such occasions by the largest, and therefore most obviously successful, respectable and knowledgeable, farmers and by the agro-allied industries. The small farmer is seldom consulted, despite the fact that those discussing his future are supposed to have his best interest at heart. Official research programs are related to the furtherance of official policies and it is difficult for research directors to find support for schemes that may have little immediate relevance to the aims of the policy-makers. If the need is to breed crop varieties that are responsive to higher applications of chemical fertilizer, and if it is official policy that in the interests of the industry as a whole more fertilizer should be used, there will be little backing for a breeding program to find varieties that grow well with smaller applications. Everyone concerned is caught on the same wheel— the government officials, their departments, the large farmers and even the bodies representing farmers who are supposed to count their constituency in numbers, rather than in acreages. So the trend toward a conglomerate system based on a few gigantic producers acquires a momentum that is difficult to check.

Over the past few years there has been considerable public pressure in the United States for information to be made available to assist farmers who wish to convert to organic methods, but departments with huge libraries of books and papers on agriculture going back for a century or more find they have little information that would be of much use. Meanwhile, the demand for organically grown food continues to increase. Is this hesitation, this reluctance to become involved, this evasiveness, an opportunity for those who practice and support the organic alternative to make a very clear pronouncement of where modern agriculture should go, how it will get there and whom it will serve when it arrives?

If it continues on its present path there are many well-informed and worried citizens of several countries who fear it may lead to a new kind of feudalism. This is how the conglomerates have been described by US Senator Adlai Stevenson III. He has spent probably more time and effort than any other US legislator in investigating the plight of farm workers, rural migration, small farmers and the relentless absorption of agricultural land, food processing and distribution into giant multinational complexes.

If this is to be our future, it may be very close. If this is to be our future it is one in which it will be impossible for the individual to farm independently. Just as the resources of the nonindustrial nations are being bought up cheaply by the multinationals, so the lands of America and Europe may be bought up by them. You and we will pay whatever price they charge for whatever food, of whatever quality, they choose to provide for us. The alternative, the organic or ecological alternative, is to scale agriculture down, to base it again on small, flexible units growing a wide range of crops for local consumption, units that offer a greater degree of stability and more responsiveness to the varied needs of the areas they feed. This alternative is to make food production serve the people who work in it and the people they feed, rather than the bank balances of stockholders many miles away. But it requires a radical change in our view of farming, of the world we live in and our proper place in it, and of ourselves.

Somewhere between these two extremes, there is a third alternative. Already our agricultural systems may have developed so far in their present direction as to make it imminent. Pursuit of the first of our alternatives will make it more probable. A change to the organic alternative may reduce the likelihood of its occurring, although for all we know it may be too late. The distinct, ominous possibility exists that within the lifetime of people already born agriculture throughout Europe and the United States may suffer a series of collapses under the strain of spiralling costs, expensive distribution, shortage of fuels and suitable sources of organic chemicals, bad technologies, biological inefficiencies, depleted reserves and lost skills.

The very existence of the organic alternative and of those who practice it provides a pool of practical experience, a new point of departure, a constant comparison and, if nothing else, a very necessary prod.

chapter V

COMMERCIAL FERTILIZERS: NO SUCH THING AS A FREE LUNCH

In Barry Commoner's *The Closing Circle,* his law of ecology says "there is no such thing as a free lunch." We know this now, or at least some of us do, but man learns slowly. Many civilizations developed extractive types of agriculture which, in the end, failed them. The Babylonians, who lived in the (once) Fertile Crescent between the Tigris and Euphrates rivers, probably supported a population density comparable to that of many modern European countries. They and the peoples of the Indus Valley and parts of Central America, the Greeks and the Romans all believed they had solved the problems of food production. They were proud of their sophisticated technology and when troubles appeared, why, technology would solve them. So it did, time and time again, until at last there was no answer. In this world you get nothing for nothing.

Farmers in Europe learned to manure their land in order, as they believed, to return to it the matter they had taken from it. The idea grew slowly and in England it was not until the late eighteenth century that most farmers were practicing a system of regular manuring. They were exhorted to do so by the agricultural societies that were formed in most counties and which spear-headed much needed social reforms and educational improvements as well as playing a major part in introducing and applying the new "scientific" approach to farming. It was in 1862 that the Earl of Derby, President of the Royal North Lancashire Society, said at the Royal Lancashire Show that "muck is money." He was urging farmers to return animal manures to the land.[1]

It was the dawning of the age of science and while the farmer soon found that if he spread manure on his land his yields increased and his crops and livestock were healthier, as yet no one knew why this should be so. In 1769, Johan Gottlieb Gahn, a Swedish scientist, discovered that a constituent of some of the manures in common use was phosphorus.

Other nutrients were identified—nitrogen, potassium and calcium. In 1840, the great German chemist Justus von Liebig, found that the phosphorus in bone meal could be made more readily available to plants if it were treated with sulphuric acid. The reaction produces gypsum (calcium sulphate) and phosphorus pentoxide, which is highly soluble in water. Von Liebig had invented "superphosphate" and the way was clear for the manufacture of the first artificial, or chemical, fertilizer.

In fact, artificial phosphate fertilizers were first produced in 1842 at Rothamsted, in England, on a farm that is now one of the country's leading government agricultural research stations. Potassium fertilizers were first manufactured in 1861, and in 1910 another German chemist, Fritz Haber, designed an industrial plant that would produce ammonia by combining hydrogen with atmospheric nitrogen. This was the basic nitrogen fertilizer and while ammonia itself has been superseded by fertilizers that are easier to use—ammonia is very volatile and subject to heavy losses by evaporation if applied to the soil surface—it is beginning to make a come-back with the development of machinery that will inject it a few inches below the surface. The usual sources of hydrogen are naphtha, a by-product of petroleum, or natural gas. The only feasible alternative to one of the fossil fuels would be sea water, but the extraction of hydrogen from water by electrolysis consumes even larger amounts of energy.[2]

These new alternatives to manure, for that is what they seemed to be, grew rapidly in popularity. In Britain in 1900, 91 per cent of the nitrogen applied to farm crops was organic in origin. By 1913 the proportion had fallen to 40 per cent.

The three main nutrients came to be known by their letter in the periodic table of elements—N for nitrogen, P for phosphorus, and K for potassium—NPK. In the year ending June, 1972, the United States consumed 41.3 million tons of NPK and in the year ending May, 1971 (the most recent year for which figures are available) the United Kingdom consumed 1.852 million nutrient tons, or 5.331 million tons of fertilizer.[3] In the most intensively farmed areas of Britain, artificial fertilizer is applied at a rate exceeding 200 lbs per acre. In the United States, NPK application rates per acre have increased each year. In 1969, the application rate for nitrogen alone, used on corn (maize), averaged 110 lbs per acre. Two hundred lbs of nitrogen became common for each acre of vegetables, much of it in the form of ammonium nitrate, the hydrogen for which came from natural gas. Within the next few years it is likely that the United States will begin importing natural gas from the USSR to meet its demand.

By 1972, US NPK application rates had begun to exceed 500 lbs per acre, with between 500 and 600 becoming the average for corn (maize) crops. In intensive vegetable growing areas, such as the Rio Grande Valley, Texas, rates reached 800 lbs per acre and more.

It is interesting to note that in 1970/71, the entire Far East region, including North Korea and North Vietnam, but excluding the People's Republic of China and Japan, consumed a total of only 4.865 million tons.[4]

Artificial fertilizers may have been introduced originally as nutrient supplements, as "tools" for increasing yields within an existing farming system that relied mainly on organic materials. However, the idea grew that N, P and K were all that a plant required. This became the official view of governments, based as it was on apparently impeccable scientific research. The words "fertilizer" and "fertility" underwent a subtle change of meaning. As applied to soils, "fertility" had always meant a large accumulation of nutrients built up and maintained over many years by careful farming. However, if plants take up nutrients from the water which dissolves them out of the soil, then it is possible to add nutrients in a soluble form so that they will enter the soil solution directly and be available to plants at once. "Fertilizer" came to refer only to water-soluble nutrients and, indeed, they were so defined. As "fertilizer" manufacturers began to acquire legislative influence in the United States, and as they came to dominate the land grant colleges through their research grants, state and federal agriculture departments established standards requiring that "fertilizers" prove to be essential to plant growth and that they be soluble in water.

At that time, the evidence appeared to be all in their favor. If a plant is raised in a nutrient solution, it will grow. Hydroponics—soilless agriculture—has received much attention and has been practiced in some countries. If plants are analyzed chemically, they are found to consist mainly of carbon (derived from atmospheric carbon dioxide), hydrogen, oxygen (derived from the air), nitrogen, phosphorus and potassium.

Impressive as it was, the evidence was nevertheless incomplete. It ignored completely the implications of an agriculture based on a total switch from organic manures and fertilizers to chemical, soluble ones. We will look in a moment at the effect this has on soils. It paid no attention to the nutritive value of the food so produced. It did not examine the long-term biological viability of such a system by growing plants in this way (hydroponically) for the ten or twenty generations necessary to show any progressive improvement or deterioration. And its analytical techniques were crude.

The usual way to analyze a plant is first to seal it in a combustion chamber that reduces it to ash. It is the ash which is analyzed. Organic compounds are destroyed, as you might expect, thus removing all the complex substances we now know to be essential. All that is left is a relatively short list of simple chemical compounds whose bonding is too strong to be broken under heat.

Of course the theory was simplistic and this was its main attraction. Remember that this was the age in which the common man "discovered" science. There was serious philosophical debate about how we would pass the time when everything in the universe was discovered and known. It was the age of reductionism. Everything could be explained in terms of "nothing but———." If there appeared a phenomenon that could not be explained, either it was purely subjective and therefore of no account, or more research would reveal its innermost secrets. The arrogance fed on itself by providing itself with apparently plausible explanations simple enough for anyone to understand. It is highly likely that scientists, whose proud "objectivity" is a myth, for they are as susceptible as any of us to emotional impulses, so planned experiments that simplistic findings would emerge. If anyone tried to defend a more holistic view of a complex world, he was dismissed as "woolly," "mystic," or "obscurantist."

So, in agriculture, a curious dichotomy began to appear. NPK farming, based on a simplistic theory, was held to be "scientific," while more advanced research was revealing that the methodology was capable of improvement. As research advanced, those working at its frontiers came to recognize that the techniques for determining whether or not a substance is essential to plant growth were inadequate and inconclusive if, indeed, they existed at all. There were investigations that showed that while certain substances may not be apparently essential to plant growth, they do play a vital role in maintaining the health of animals and, in the long term, they do assist plants to grow more prolifically.[5] It was discovered that essential nutrients locked in insoluble materials will be made soluble, and so released to plants, by the action of microorganisms. This should have surprised no one, for unless it is so, it is difficult to see how *any* plant growth would have been possible before the introduction of chemical, soluble fertilizers. However, it showed, too, that nutrients released in this way, slowly, are less likely to be lost from the soil by leaching.[6]

Yet the strangest paradox is still to come. Having assured themselves that plants require "nothing but" N, P and K, scientists went on to discover that an average acre of farm soil contains between 1,500 and 7,000 lbs of nitrogen, with an additional bonus of 35,000 tons in the air above it, between 750 and 3,500 lbs of phosphorus and probably close to 40,000 lbs of potassium, all within plough depth. The response of the agricultural scientists to this startling piece of information was *to urge farmers to buy more NPK fertilizer!* They even went so far as to advance the view that the soil was of such little importance that, provided an adequate supply of fertilizer was available, it could be regarded as no more than an inert substrate. The land was important to agriculture simply as a convenient place to put the plants while they are growing, and the soil was useful only to prevent them from falling over or blowing or washing away! So successful were they in convincing governments and the farmers themselves of this extraordinary piece of double-think that when the British government held an inquiry into the effects of farming on soils, the committee was able to consider soil structure and fertility as separate issues. While they found serious deterioration in the structure of many British soils, and very low levels of organic matter, they were able, nevertheless, to reassure the Minister that:

"As far as the *nutrient fertility* of our soils is concerned, we have few misgivings. There is no evidence to show that the disappearance of livestock from certain areas and the replacement of ley-farming (pasture-farming) and farmyard manure by chemical fertilizers has led to any loss of inherent fertility. Nor is there any evidence that organic matter is intrinsically a better source of nutrients."[7]

If organic matter is being compared with soluble fertilizers on the basis of its content of soluble nutrients and provided chemical fertilizers are available, all is well.

We hope we may be forgiven for suggesting that a more appropriate line of investigation for agricultural scientists would have been to discover ways of utilizing more efficiently those nutrients provided free, by nature and the wise farmer, but we accept that this would be naive. The direction of agricultural research has been determined by commercial and political influences, rather than by common sense or the "impartial search for knowledge for the benefit of man" in which so many scientists still believe they are engaged.

Officially and academically, then, farmers were encouraged and urged to turn their backs on skills and knowledge, much of it intuitive, acquired over thousands of years and to reject the fer-tilizing materials that had always been available on or near to their farms. Of course, the lure was attractive. NPK fertilizers were much easier to handle than bulky organic manures. They were delivered to the farm in neat plastic bags. The manufacturers of farm machinery went along with this development. They closed their minds to the alternative need to produce equipment to enable farmers to handle their bulky materials more easily and so to "modernize" traditional methods of recycling. It would not have been difficult. Much of the machinery needed no more than modification and where the technology did not exist it was not so far removed from what did that it could not have been devel-oped. As it is, many organic farmers have solved such handling problems themselves, but they have received little help or en-couragement from the industry whose equipment they use.

So the race was on. Farmers were intensifying their farming in the hope of intensifying their profits. As they did so, they found that capital and operating costs intensified as well. This led them to seek economies, ways of cutting corners. *NPK became a substitute for land*! In Europe, which is crowded, there might be some excuse for such a substitution, at least until it was dis-covered that it could not be made to work indefinitely, but in the United States, which has a low population density and thus a plentiful supply of land, the absurdity is obvious and the result is predictable. As the volume of additives used per acre increased, and as modern farm machinery, irrigation and hybrid seeds were introduced, millions of acres, millions of farmers and farm work-ers were withdrawn from production or forced to leave agriculture —and their homes and way of life. In the vacuum they have left, the legitimate farmers compete among themselves and against tax-favored speculators and corporate newcomers to be "big enough" to stay in business.

In one way or another, in most of the industrial countries, artificial fertilizers are supplied to farmers at a subsidized price. The subsidy may be direct, in the form of a refund of part of the price paid as it was in Britain until entry to the EEC, or hidden in tax reliefs, or in the artificially low price of energy. The fertilizer industry is linked closely with the petrochemical and explosives industries, both of which are of military importance and neither of which are governments prepared to see become depressed during peace time. Yet as political tensions, within Europe at any rate, are relaxing and, anyway, military requirements have changed, fertilizer prices may begin to rise to more realistic levels. Indeed, as the energy required to produce them increases in price, fertilizer costs must rise.

We admit that NPK nutrients, combined with modern farm equipment, irrigation, and hybrid seeds have produced impressive results. Yields rose dramatically. In the nineteenth century, the average yield per acre for wheat in Britain was 16 cwts. By 1940 it had reached 19 cwts but by the early 1960s it reached about 32 cwts, a level it has sustained, more or less, ever since.[8] Other crops showed comparable gains. It was hardly surprising that farmers connected the increases with fertilizer use, which rose most sharply in the period from 1945 to 1960. These were the boom years for farmers and although yields levelled off early in the 1960s, for many years governments and farmers themselves have believed they would pick up again.

Almost certainly they are wrong. Throughout the 1960s and into the 1970s in the United States, and particularly in areas of intensive production, such as those in the state of California and in the Rio Grande Valley of Texas, "decline" has become a major concern—walnut decline, prune decline, olive decline, "California citrus decline," declining yields and quality in general as more land fails to produce as much, and as more land becomes polluted with salts and infested with destructive soil organisms. Although fertilizer consumption continues to rise and farmers are still being urged to "intensify their farming" with closer cropping and more NPK additives, yields have not always increased and in some cases they have continued to fall still lower.

Pests and weeds have become more destructive in both countries than they were for many years. The structure of some soils is deteriorating. Minor dust storms have blown away topsoil in Great Britain. Several inquiries by the farmers led to an official investigation which found that soil structures were, indeed, deteriorating, and that undoubtedly this was due to modern farming practices. Yet, the Agricultural Advisory Council did not connect this deterioration with the use of artificial fertilizers, nor did they admit of any problem regarding the fertility of soils.

In the United States, the official, public position has been to pretend that all is better than ever, and to deny that any "significant" soil problems exist.

As the use of artificial fertilizer has increased, less and less organic matter has been returned to the soil, while reserves of humus have been mined away. Farmers often believed that applications of NPK additives were an adequate substitute for the nutrients removed, but their view was based on a theory which is far from complete.

The first flaw in the theory is that it overlooks entirely the contribution organic matter makes to soil structure. Plants take up nutrients from an aqueous solution and the nutrients are released into the water by the action of microorganisms which require oxygen. So a fertile soil must contain water and oxygen and both must be able to move freely through it. Water must move first downwards from the surface, as it drains, and then upwards again by capillary attraction. The soil must have channels large enough to permit drainage and small enough to have a capillary effect, and there must be spaces between soil particles which contain air. This is what is meant by the structure of a soil and this is why it is essential. Even though analysis may show that a soil is rich in nutrients, if it lacks structure the nutrients will not reach the plant roots and the rate at which they will be made available will be very slow.

In most, but not all, soils, the structure is provided by decaying organic matter, and by the soil organisms, large and small, which break it down into relatively simple chemical compounds.

Imagine the fate of a leaf lying on the surface. First it is attacked by insects, mites and fungi, which consume part of it and break up the remainder into small pieces. These are carried into the soil by earthworms. As the earthworms move through the soil, they make tunnels, which are left behind them. Some of the leaf they will eat, some they will leave. As the small pieces decay by the action of bacteria, they, too, leave behind them the space they once occupied and this space is now lined by the colloidal substances secreted by the bacterial populations. These help to consolidate it.

The numbers involved in this process are truly astronomical. Imagine, if you will, a pine needle 60 mm $\times$ 1 mm $\times$ 0.5 mm in size. The surface area of this needle is 180 sq. mm. Small soil animals may cut it into 60 pieces, each 1 mm thick, so increasing its surface area to 240 sq. mm. There is then a range of organisms that can break these particles into cubes with 10 micron sides. This produces 30 million fragments with a total surface area of 18,000 sq. mm. A nematode may then cut these pieces into still smaller ones, with 0.1 micron sides. This gives us 3 billion pieces and a total surface area of 1,800,000 sq. mm, or 1.8 sq. meters— and all from a 60 mm pine needle.

This process takes place constantly, day in, day out, and it gives some idea of the complexity of the soil as a living eco-system in which the mineral soil particles are pushed aside and mixed with organic matter to give the soil the crumblike structure farmers and gardeners recognize.

The size of the living populations of the top few inches of soil may surprise you. They are difficult to count and such figures as we have are probably on the low side, but French workers examining the soil of an oak forest near Paris counted the populations of protozoa, nematodes, arachnidae and insects, taking *one* species of each. The protozoa live at a density of several hundred million to the square meter, 10 cms deep. There were more than 30 million nematodes and 45,000 *Acaridae,* the species of arachnidae they chose to count. The insect they selected, the *Collembola,* has between eight and nine thousand known forms. They did not report on its numbers, but Sir E. John Russell, in *The World of the Soil,* said there may be 248 million per acre in the top 12 inches of soil. If you see a field with cattle grazing, the microorganisms in the topsoil of that field probably weigh more than do the cattle.

The second flaw relates to the nutrient theory itself. In the early days, four main nutrients for plants were identified—NPK and calcium, which is usually supplied in the form of lime. Later we learned that there is also a range of metals that are necessary in tiny amounts, measured in parts per million. Sometimes these are added to artificial fertilizers, but without adding much benefit to them. There is good evidence that the heavy use of N, P and K inhibits the uptake of at least some of them. Superphosphate, for example, inhibits the uptake of zinc, copper and iron. Deficiencies in trace minerals, as they are called, may show in the plant, but if they are not visible they may lead to deficiencies in the consumers of those plants. Trace mineral deficiencies have been detected in unprocessed rations fed to livestock in zoos, laboratories and on farms.

If NPK fertilizers are compared with organic fertilizers, the effect on trace mineral uptake is a multiplicative one, for while NPK inhibits uptake so effectively, reducing the amounts available, organic manures and fertilizers add trace minerals without inhibiting uptake, so effectively adding to the amounts available.

It is not known with any certainty why NPK should inhibit trace mineral uptake, but one theory which might account for the phenomenon suggests that the soil may be seriously, but subtly, unbalanced. In the soil solution, nutrients exist in anionic and cationic form, as small molecules carrying a positive or negative electrical charge. The outer membrane of the cells of the root hair is also charged and so it attracts molecules bearing the opposite charge. Each time one passes through the membrane, its own charge is reduced until it becomes neutral, when it will take up no more nutrients. Now if concentrated doses of particular nutrients are applied in the region of the root hair, the number of trace mineral ions and cations as a proportion of the whole will be reduced. In effect, they are being diluted, and since the mechanism by which they enter the plant is only partially selective, and since there is an upward limit to the total amount that can be taken up at any one time, the root is likely to receive fewer trace minerals.

The situation may be remedied by adding trace minerals as such, but this may lead to an accumulation in the soil. Most trace minerals are essential in small doses and highly toxic in large doses and no one knows where the dividing line is. So simply adding more may appear to restore the balance, but it may do so at the price of severe damage to the future capacity of the soil to produce food safely and abundantly. Of course, if the problem is caused by the inhibition of uptake, rather than by simple deficiency, the imbalance may be compounded, with little beneficial effect.

By relying on organic manures and fertilizers, organic farmers are not likely to suffer from trace mineral deficiencies unless these are inherent in their particular soils and even then, by utilizing organic waste materials which may have been produced in other soils, regional deficiencies can gradually be ironed out, and safely. Consider, for example, the huge volume of trace minerals, vitamins, protein and other nutrients vital to plant growth and human health which are lost permanently, for all practical purposes, when millions of tons of manure and garbage are hauled annually out to sea or packed in landfills. Organic matter is nature's great storehouse. It is the only material on our small planet that will store the sun's energy.[9] By locking up carbon in large quantities, it preserves an atmosphere with adequate oxygen. Trace minerals, major minerals, and other nutrients, already processed and refined naturally, are held in organic matter and released to become available again through decomposition.[10] Investigations show consistently that soils high in humus and organic matter are correspondingly high in trace minerals, and that plants grown in such soils produce foods with high and balanced trace minerals.[11]

Good evidence exists already which suggests that biotic, or organic, compounds, produced by the synthesizing action of soil microorganisms and the concentration of humus, are essential to healthy plant growth and to the uptake of nutrients essential to plants and to animals. When humic acid, for example, is applied to soils which are low in humus content, plant growth is stimulated and production is increased. Investigation has also shown that humic acid facilitates the uptake of zinc and will correct chlorosis caused by high salt content.[12] Moreover, humic acid content in the soil increases in proportion to the percentage of humus in the soil.[13]

Studies begun in the early 1970s have already confirmed the long-held organic view that most plants, under normal field conditions, are dependent upon an association with certain soil microorganisms for the adequate uptake of minerals and other nutrients, organic compounds and moisture. Colonies of organisms like the fungus mycorrhiza, once described by Sir Albert Howard as the "key" to the wheel of life, attach themselves to roots of plants and, in a manner of speaking, become the plant's feeders, drawing in and transmitting vital substances from the surrounding soil.[14] One interesting study that compared the "sucking power" of the plant root with that of the associated fungus found the power of the fungus to be many times the greater.[15] Numerous investigations have shown that mycorrhizae and colonies of other advantageous soil microorganisms increase as the humus content is increased.

Leading from this, the third flaw is that the fertilizer-nutrient theory was developed before anything was known of the role of soil organisms in plant nutrition, or of the effect of the heavy use of NPK on soil micropopulations. Soon after manuring became common practice, it was observed that the growing of certain crops improved the fertility of the soil much as manures did. Eventually it was discovered that these plants, legumes, have nodules on their roots which contain colonies of bacteria that are able to "fix" atmospheric nitrogen. That discovery was made a long time ago. Now we know that there are many nitrogen-fixing organisms in the soil and that even at modern fertilizer application rates, by far the largest part of the nitrogen taken up by plants is provided naturally, by the populations of organisms in the soil, either directly from the atmosphere or from the decay products of organic matter. However, it is likely that in intensively farmed areas, the quantities of nitrogen added by man now represent a significant proportion of the total nitrogen engaged in the nitrogen cycle and ecologists fear that the cycle may be gravely disrupted.

The same is true of the other nutrients. In a natural, "wild" soil, nutrients are being cycled all the time from the soil, through plants, herbivores, omnivores, carnivores and back to the soil again. Losses are made good by supplements of nutrients provided by autotrophic organisms from the air and from the rocks, the purely mineral content of the soil. All of the cycling is performed by living organisms, for which the substances they cycle are either food or waste product. Clearly, if we add concentrated doses of similar substances to the soil, we are bound to alter the balance of populations, and it is impossible that such a change can be advantageous to the farmer. Those organisms to which this addition represents a food supply will increase in number, so locking up the nutrient, which is not what the farmer intended. Others, to which it represents a waste product, will decrease in number, since it is a basic law of biology that the waste products of any organism are toxic to that organism. Yet it is this group the farmer would prefer to see increase, for it is they which release nutrient to the plants. There are still some scientists who maintain that artificial fertilizers have no effect on soil populations, but in most soils, under most conditions, it is unlikely that they can be right. Logic alone would suggest otherwise. The uncertainty arises from the fact that even today, as we have seen, we know comparatively little about soil ecology, which seems far more complex than it was imagined to be even a few years ago.

It is because of this disruptive effect on micropopulation balances that the fear exists that nutrient cycles may be disturbed. If the populations that are reduced are those which are essential to some stage in a cycle—and this applies to most soil organisms —there must be a risk that the cycle will be slowed or arrested at that stage. If this is so, then each time he uses artificial fertilizer, the farmer will be taking over the role of those organisms and if the fertility of his soil is to be maintained he will have to go on performing this essential function he has made for himself. In ecological terms, he has displaced one set of organisms and has occupied their niche. The system which once depended on them now depends on him. Perhaps this would not matter too much were the farmer able to occupy this niche as efficiently as those organisms he displaced, but unfortunately he does not, for all the reasons we have mentioned above, but also because he cannot possibly afford to provide nutrients in the quantities that would be required by a fertile, active soil.

If we are right, we should expect to find that farmers who use large quantities of fertilizers have to increase their rate of application over a period of time in order to sustain constant yields. We do find this to be the case, although the concomitant deterioration in soil structure also contributes.

At best the situation is ludicrous. The soil provides its nutrients free of charge. When the farmer begins to supply them himself, it costs him money and he must balance the expected increase in yield against the cost of the fertilizer. Again, if we are right and the more he applies the more he will *have* to apply in the future, then the day must come when he will experience diminishing and then negative returns. That day came some years ago.

Already the most progressive, even of the "orthodox" farmers, are seeking to reduce their fertilizer costs. Longwood Estates, in Hampshire, England, which owns more than 10,000 acres of chalky downland and has access to almost unlimited capital and whose farming is computer-monitored, field by field, has decided that the most sensible, and profitable, kind of farming under their conditions calls for a reduction in fertilizer use. They plan to establish over the whole area a five year rotation of four years grass and one year arable and to raise dairy and beef cattle on grass. They have rejected intensive indoor rearing and feed lots as uneconomic, although the cattle have to be wintered in yards, as they do on most British farms. All the manure will be returned to the land and the fertilizer bill brought down to a minimum.

In the United States, farmers large and small are quietly, and sometimes in desperation, incorporating organic soil management practices. After years of NPK farming and declining yields, one major US farm corporation was faced with the prospect of either going out of business or improving its soil. A well-known agronomist was hired and after five years a building up and balancing the soil by additions of trace minerals, rotations and heavy cover-cropping to add organic matter, profits increased by $371 per acre. This demonstrates again that as farmers come to rely more and more on NPK additives, and their use intensifies, the cost will come to exceed the initial boom they produced. This same agronomist is now working with the Davis Ranch Corporation in Lompoc, California, and already he has placed a part of their acreage under organic soil management as together they investigate ways to overcome high salinity, poor friability, declining quality and low humus content.

Humus aids soil structure, but its main value lies in its chemical composition. This is extremely complex. All the constituents of organic matter do not decompose at the same rate. Those that are soluble in water, together with the celluloses and hemicelluloses, are removed fairly quickly, while at the same time there is an increase in the content of lignin and lignin complexes and of proteins. No one knows just why the protein content should increase, since proteins added to the soil degrade rapidly, but it does. Possibly the protein molecules are adsorbed onto the surface of clay minerals and so rendered resistant to further decomposition. Since nitrogen is a prime constituent of proteins, humus is able to store nitrogen and to release it slowly over a prolonged period of time. It is a kind of fertility bank, as we mentioned earlier. The nitrogen content is usually between 3 and 6 per cent.

It is not only nitrogen that is stored in humus, however. As organic matter decays, chemical changes take place that increase its cation-exchanging capacity. This is a technical way of saying that it is able to hold certain molecules with which it comes into contact. So it traps and stores a range of nutrients, releasing all of them slowly.

Humus is highly insoluble in water, but it absorbs large amounts of water by swelling and then shrinking again. So it helps to retain moisture in the soil.

There has been much discussion of the optimum content of humus, or of total organic matter, in soils. The figure of 3 per cent has been suggested widely, but the truth is that each soil, in each climate, has its own optimum and there is wide variation. In 1953, it was estimated that the soils of the United States contained an average of 3 to 6 per cent humus and it is believed that the original, virgin soils contained something like 4.25 per cent. In Britain, where it is quite impossible even to estimate the types of the original soils, desired organic matter levels have been decreased through the years. At one time farmers were advised that their soils should contain 5 per cent, then the figure was reduced to 3 per cent. Today, the official view is that soils with less than 1 per cent give cause for concern. As it is, there are many soils which continue to give good yields with little more than 1 per cent, for the time being at any rate. Only time will tell whether such a low level can provide a stable, fertile farm soil for very long.

The argument about percentages is at best pointless and at worst dangerous, for it leaves the way clear for the agrochemical industry to reassure farmers that the organic content of their soils is of no importance whatever. Farmers are not fools, but they are as prone as the rest of us to accept the view which approximates most closely to what they would like to believe and if experts cannot agree, then this is just what many of them will do. Let's face it, it would be much more convenient for all of us if the world were the simple place our departments of agriculture and agrochemical salesmen would have us believe it is!

While fertilizer practices, combined with a reduction in the amount of organic matter returned to the soil, tend to reduce the soil's humus content, humus can also be removed by erosion. It has been estimated that in the United States, farms and grasslands lose three billion tons of topsoil every year by erosion, mainly caused by man; and today, after 35 years of service from the US Soil Conservation Service, nearly two-thirds of the nation's 1.5 billion acres of privately owned rural land is in need of conservation treatment. A single windstorm has been known to remove as much as 150 tons of soil from each acre of land.

Erosion is accelerated by the practice of leaving the land bare during the winter. This has become increasingly common, both in Europe and in the United States and the removal of livestock from Eastern England has made it practicable to clear away the hedgerows once used to retain and shelter animals. This has released more land for cropping and has eliminated maintenance costs.[16] Hedgerow removal, which once qualified for a government grant because it was said to improve the farm, is no longer actively encouraged, and today there is more interest among farmers in planting shelter belts of trees to protect exposed fields.

Even so, for a decade or more, hedges were ripped out at the rate of between 5,000 and 10,000 miles a year and although the rate of removal has slowed down, each year a few more miles disappear.

The increase in fertilizer use during a period of static or declining yields suggests that surplus fertilizer has been lost, but by a basic law of ecology, it must go somewhere. There is a limit to the amount of nutrient that a plant will take up. While some of the increase in fertilizer use has been made necessary as the farmer has, to some extent, taken over the role of some groups of soil organisms, much more is due to the loss of humus and structure. It has become necessary to apply more fertilizer simply to ensure that the same amount reaches the root system. Large amounts are lost.

This has led in places to problems of eutrophication and of the contamination of water supplies. A few years ago, "eutrophication" became something of a vogue word and there were highly colored and sometimes rather overdramatized stories of the Great Lakes dying. Like every other ecological problem, the truth is more complicated.

Soluble nutrients leach from all fertile land whenever it rains. They move through the ground water, or in surface runoff, until they reach rivers and lakes, where, because they stimulate plant growth, they make it possible for ecosystems to exist in water, in exactly the same way that they support land-based life. Surplus nutrients, together with plant debris which decomposes to return its nutrients to the cycle, tend to wash into lakes and so over a period of time there is an accumulation of plants in lakes. The distance they extend from the shore will depend, obviously, on the depth of the water, since most of them are photosynthetic. A point will be reached when plant production exceeds the rate of consumption by herbivorous aquatic animals and as the surplus decomposes there will be an increase in the population of the bacteria engaged in this operation. These bacteria consume oxygen and so the stock of free oxygen in the water decreases. This, in turn, kills higher life forms by asphyxiation, so reducing still further the rate of consumption; thus, the downward spiral accelerates until the lake is reduced to a swamp and eventually dries out completely. The drying out itself is accelerated by the increase in plant growth, since plants release large amounts of water vapor into the air by transpiration. So it is the natural fate of all lakes to "die" of old age and to disappear. In nature this process takes thousands, or sometimes millions, of years.

Since chemical fertilizers are added to the land in highly soluble forms, it is inevitable that they are leached out quickly and so it is obvious that the nutrient content of water systems will increase. The effect on slow-moving rivers and lakes is to speed up dramatically the natural aging process. It would be unfair, however, to place the whole of the blame for this subtle form of water pollution onto the shoulders of the farmers. The main nutrient source in most systems is still sewage, which contains nitrogen and phosphorus, and detergents, which contain phosphorus.

Ground water itself may become contaminated and it is not uncommon for nitrates and phosphates to enter domestic water supplies. Conventional water treatment plants are not equipped for the complex and very expensive third stage of water purification, which is necessary to remove plant nutrients. There is much heated discussion about the "maximum safe dose" of nitrates in potable water and in parts of lowland England they have sometimes reached near-danger levels to infants, who may suffer from methemoglobinemia. In parts of the United States, too, toxic levels have been reached occasionally, and nitrogen pollution has become a serious problem in, for example, Illinois.

Dare we point out, yet again, that this represents a serious loss of nutrient that could, and should, be used?

The loss of organic matter also affects drainage patterns. As structure is lost, water may run off the surface of land, rather than draining through it vertically. Where cattle stocking rates are too high, or where heavy machinery has been used at the wrong time of year, compaction is more likely and water may accumulate above a hard, impervious pan. The result may be recurring cycles of flood and drought and the response, particularly in arid areas where the water-holding capacity of the soil is of great importance, may be to increase irrigation. This is a more complicated practice than it may appear and it can be double-edged. Unless the structure of the soil is improved, irrigation may simply accelerate the rate of erosion. Americans spend more than 250 billion dollars annually in dredging topsoil from reservoirs and navigable waterways and it is estimated that 2,000 irrigation dams in the United States are now completely silted up. It was the failure of their irrigation systems that caused the collapse of the farming systems of the Fertile Crescent and of the Indus Valley.

The difference in the water-holding capacity of soils with a high humus content as the result of years of careful organic farming are very evident whenever the weather is bad. After wet winters, organic farmers often find they can move their cattle out of doors earlier than their neighbors, because their soil has a better structure and can stand up to the weight of the animals. When it is dry, organic farmers find their pasture remains greener for longer than does that of their neighbors. At the Soil Association farms it was observed that the same tractor ploughing fields that differed only in the treatment they had received, experimentally, over the years had to work longer in the fields with a lower organic matter content than in those which had been farmed organically. The difference was measurable in the amount of fuel used.

If a farmer has a problem he would be well advised to seek advice elsewhere than from a fertilizer company or agency. If his problem arises through the lack of humus in his soil, the advice he is given, which will be to increase his rate of fertilizer application, is likely to exacerbate the situation. Had he approached an independent soil scientist he might have received very different advice. He might have been told to buy, beg, or, if necessary, steal organic matter.

In spite of all the advice they have received to the contrary, and in spite of all the "scientific" articles they have read in their farming magazines, many farmers have been taking from the soil more than they were returning to it for years. The decline in their yields, the deterioration in the quality of their produce, the increase in weeds, pests and diseases from which their crops suffer are all part of the price they must pay.

As we have seen, there is a clear and direct relationship between the amount of organic matter returned to a soil and its humus content. One of the criticisms levelled at those who advocate the organic alternative is that there is simply not enough organic matter to go round. How could present levels of production be maintained, they ask, when those farmers who do use organic matter often have difficulty in locating adequate supplies?

On the face of it, this is true, but a closer examination of the nutrient economics involved reveals the fallacy. The problem is not that the organic matter does not exist, but that it is in the wrong place. The food produced on our farms is consumed by livestock or by humans. As the livestock have become dissociated from arable farming, the wastes they produce accumulate close to the unit and far from the land that grew their feed. As urbanization continues, more and more of the human consumers live a long way from the land and so their wastes accumulate as well. So we have two problems which should be solved together—a shortage of organic matter for our farms and an accumulation of organic matter in our towns and cities and close to our livestock units. The remedy is in our hands, as we have seen. The technology exists and all that stands in the way of its being applied is an economic barrier. Surely, the economics must be wrong. Surely, since economic systems are entirely man-made, they can be modified. Is it not a question of priorities, after all?

Even the problem of toxic industrial wastes in sewage can be overcome. Tolerance levels for the discharge of heavy metals should be reduced, either by a form of direct penalization, or through a resource tax which encourages more careful use of them. This would reduce the quantities of toxic material that need to be dealt with. Industrial waste should be kept separate from domestic sewage. This would ensure that such toxic materials as are discharged do not become associated with the organic matter we wish to recycle and at the same time it would make the toxic materials themselves easier to recover.

In fact, the problem itself may have been exaggerated. Experiments in which plants have been grown in soil treated with municipal composts from a number of towns applied at rates of 200 tons an acre or more revealed little evidence of uptake by plants, which were quite safe to eat. If municipal composts made from currently available wastes by present techniques are used in moderate amounts, it is unlikely that any serious danger exists. Sewage, however, is another matter and sewage from industrial towns probably should be composted with other wastes before use in order to dilute toxic substances.

Unless we learn to use more efficiently the resources at our disposal, so conserving those that are in short supply, we may suffer serious reductions in our standard of living in the not-too-distant future. The soil is our prime resource, together with water and air. At present the majority of our farmers, pressured by governments and by an aggressive agrochemical industry, are not using their soils efficiently. They are farming extractively, mining the soil, and the consequences are beginning to be felt. You cannot take from the barrel more than you put into it.

In a way it is possible to look upon the requirements of good health as the contents of the barrel, and this might make it easier for us to realize that the nutrients that are vital to good health originate in the soil and in our management of it. The farmer who sets out to produce organic food and who adopts farming practices with this end in view is more likely to achieve his aim than one who is concerned only with producing the largest bulk at the lowest cost. This raises again the vexed question of differences in quality between food grown organically and that grown non-organically. It is not possible to give a final answer, but it is possible to summarize a few of the facts that are known and that have been reported by agricultural scientists in the United States, in Britain and in Europe.

The mineral quality of soil influences the mineral balance of food grown in that soil.[17]

The nutritional values of foods produced vary according to geographical areas and the soils in which they were grown.[18]

In general, application of large amounts of nitrogen will decrease the calcium and phosphorus content of plants.[19]

Foods produced using organic fertilizers show higher nutritional value than foods produced using NPK plant food supplements.[20]

Excessive applications of plant nutrients result in excessive consumption of the nutrients by the plants.[21]

Applications of trace minerals to the soil with NPK nutrient supplements are inconsequential.[22]

Superphosphates can inhibit the uptake of zinc, copper and iron by plants.[23]

Applications of nitrogen have been shown to result in reduced iron content in the root tissue of plants.[24]

The heavy and increasing use of nitrogen-containing chemical fertilizers has resulted in nitrogen compounds entering municipal water supplies, lakes and rivers and has increased the nitrate content of plant materials.[25]

Nitrate nitrogen in water above 10 or more milligrams per liter (parts per million) has been shown to be hazardous and even fatal to infants.[26]

Herbicides, fungicides and fumigants destroy essential soil microorganisms, particularly mycorrhizae. Decorative and citrus plants produced in sterilized nonmycorrhizal soil have been shown to require toxic amounts of NPK plant food additives in order to produce the same vigorous growth achieved in mycorrhizal soils using small amounts of nutrients.[27]

Fertile soils high in humus and organic matter show correspondingly higher ratios of vitamins, proteins and antibiotics and these have been identified in plants produced in such soils. Plants not grown in fertile soils respond positively when organic substances found in fertile soils are introduced.[28]

chapter VI

CHASING OUR TAILS

It can be shown, on purely theoretical grounds based on general principles, that the use of pesticides is bound to be self-defeating provided only that sufficiently large quantities are used, that they are sufficiently toxic and that their use continues long enough. As a central part of any farming system their routine use makes no sense, although there may be particular instances in which the farmer has little choice but to use one to control a situation that has deteriorated too far to be controlled in any other way. In such a case, pesticides are available that will prove adequate and that have none of the undesirable side-effects which have earned the modern synthetic products such notoriety.

Let us imagine an area of land on which a farm crop is growing. In addition to the crop there is a wide variety of other plants among the crop plants and around the edges of the crop area. These plants support a similarly wide variety of insect species and, as we have seen, the soil supports the most varied populations of all. All of these plants, insects, microorganisms and most of the larger animals represent a source of food to others and most of them breed rapidly. Should the population of any one species increase there will be a proportional increase in the populations which feed upon it as they are presented with an apparently limitless food supply. The numbers of the original species will be reduced and so will that of its enemies as the food is consumed. A balance will be restored. However, should any particular population increase too rapidly and assume too large a size to be controlled quickly and effectively, the farmer may have a pest problem and he may decide to use a pesticide. What happens then is best illustrated by the use of insecticides, but to some extent it applies to herbicides and fungicides as well.

The insecticide does not discriminate. It will kill any insect with which it comes into contact. Thus it kills not only the pest, but also the enemies of the pest. Since these may breed more slowly than the pest and since their development is bound to follow behind that of the pest and, in any case, their numbers are smaller, the effect on their population may be more severe than the effect on the pest population itself. At best their ability to control the pest will be diminished and at worst they may be so decimated that they die out altogether. So, although a large proportion of the pest population may have been eliminated, the farmer will also have eliminated many other insects he should regard as allies. The probability of a recurrence of the pest problem will have increased.

The reduction of the pest population to a manageable size has not removed the unlimited food source—the crop—which led to its appearance in the first place. Of course, the crop will not remain in the field for very long, but unless it is removed quickly, there is a possibility that the reduction in the numbers of one species may have left the food supply available to a rival species. One pest will simply be exchanged for another.

Populations are adaptive to changes in their environment and their rate of adaptability is a function of their rate of reproduction. The more rapidly they reproduce, the more rapidly will they adapt. If a poisonous substance is introduced into their environment it is only a matter of time before they learn to tolerate it. The species most frequently exposed will adapt most quickly and the insecticide will lose its effectiveness. In the case of insects, this may happen in one of two ways at least. Certain insects will possess a particularly thick chitinous outer covering that is impervious, so that the poison cannot enter their system. Others will possess the ability to detoxify the poison and eliminate it from their bodies. Such individuals will survive and since these characteristics are

Organic farming has also been termed ecological farming because it embraces the idea of working with natural processes for the long-term benefit of all living things. Certainly steers are happier, healthier animals when they are raised in a natural environment and not forced into feed lots where they are treated as potential dollars, not as living animals. And men and women are better off eating meat fattened in nature's own time, without any growth stimulating hormones or potentially dangerous antibiotic and pesticide residues.

transmitted genetically, they will breed them into their progeny. Indeed, they may breed more rapidly than their fellows, since the weaker members of the species, who do not have this particular tolerance, will be destroyed. So a resistant population will develop by natural selection in response to a stress introduced by man.

This is not an argument against the occasional, discrete use of a pesticide, but it is an argument against routine pesticide applications. In the long run they are bound to prove disruptive, unbalancing still further the system the farmer seeks to stabilize, and increasingly ineffectual.

If this is so, why are pesticides used so widely? In the course of this book we have had a good deal to say about energy, and pesticide use, too, bears on this.

An ecosystem exists in one of two stages; it may be a successional or a climax system. In a successional system, species are arriving to fill niches available to them and the range of species the system supports is increasing. Once the system reaches its climax there will be room for no more newcomers. Every possible source of food will be providing energy for a living organism and there will be very little that is wasted. Such a system may remain intact and stable for a very long period of time.

When farmers cleared land to plant their crops, they were removing a complex, climax system, and substituting a simpler system dominated by the limited number of plant species they wished to grow. Ecologically, they changed a climax system to a successional system or, if the area into which they moved had not yet reached its climax, they substituted one successional system for another. As a result, new species began to arrive to fill the niches created. Species which regarded the crop plants as food, predators which preyed on them, plants which were able to compete successfully with the crop for sunlight or nutrient all began to appear and the farmer found he had pests and weeds. He had competitors for his food. It was inevitable and weeds and pests are an inevitable result of farming as we understand it.

The entire system derives its energy from the sun. It was this energy, too, which stimulated the growth of the pests and weeds

as well as of the crop. If the farmer was to overcome his competitors he had to apply energy and, moreover, he had to apply
more energy than was being supplied by the sun. This was easier
in some regions than in others. In the temperate climates of north-
ern Europe, there is less solar energy available over the year than
there is in the tropics. So farming was easier in these climates.
Furthermore, these regions enjoyed—if that is the word—a winter
in which temperatures fell below the minimum required for plant
growth and so helped to slow down the development of weeds
and, with them, pests. Of course, it prevented farmers from growing crops as well, but in this case they were able to adapt and
base their farming patterns on the seasonal variation. In some
tropical countries, too, there was a seasonal restriction in plant
growth, in this case provided by a pronounced dry season when
plants could not grow for lack of water. This had an effect similar
to that of the temperate winter. In the wet tropics, on the other
hand, it was far more difficult to check the unwanted proliferation of species and for this reason the kind of agriculture
practiced in Europe and North America has never worked well
in those areas and is not appropriate for them.

In the early days the energy the farmer supplied was his own
and that of his family, aided by simple implements. Later he
learned to use draught animals and, later still, machines. These
all increased the amount of energy he was able to expend on the
control of his ecosystem. This is what environmental management
means when applied to farming.

To a large extent, farmers supplemented the brute force available to them by a mixture of acceptance and skill. They accepted
that certain losses were unavoidable and so they allowed for
them, just as they allowed for the fact that not all of the seed
they sowed would germinate. There is a very old English farming
saying: "One to plant and one to grow, one for the farmer and
one for the crow." They accepted that a field completely free
from weeds was as impossible to achieve as it was unnatural.
They grew up with and enjoyed the wild flowers and the insects
and yet they managed to keep both within acceptable limits.

If the farmer waits until the weeds in his field germinate before he sows his seed, he can kill most of the weeds with his cultivations and at the same time he can know, because seeds have germinated there, that the ground is ready for his crop. If he observes that certain insects multiply rapidly at a particular stage in their life cycle, a delay in the sowing of his crop may leave them without a food supply when they need it most. There are a thousand ways in which the farmer learned to live with nature. He achieved a balance, and the world was fed.

Pesticides represented a new source of energy. Indeed, a great deal of energy is consumed in their production and most of the modern pesticides are derived from petroleum. They changed everything. They promised to rid our farms of weeds and pests completely and permanently. This they would do by poisoning every species that did not contribute directly to the production of food. A new dimension had been introduced into plant protection. Farmers need no longer try to work with nature, often using its own methods to their advantage. The principle now was to override nature altogether by force—or, if you prefer, energy. At the same time, pesticides promised to make farming possible with less skill. After all, it requires little enough skill to operate a sprayer.

We saw with the use of chemical fertilizers that once the farmer began to use them extensively he was creating a situation of dependence and he left himself with no choice but to go on, probably increasing the rate of application over a period of time. Were he to abandon them suddenly and completely, he would suffer serious losses in yield until his soil had restabilized. This is the position with regard to pesticides. Once the farmer comes to rely on them he is caught. He has reduced the capacity of the community of plants and animals on his farm to maintain its own stability. Moreover, the belief that pesticides gave him more power than in fact they did encouraged him to overreact. Many of the acute, short-term problems that pesticides have created are the result of the "overkill" mentality that led farmers to regard all insects as pests and all noncrop plants as weeds. There have been countless cases of sprays being used against nonexistent pests and weeds.

In fact, such power never existed. It was never possible to eliminate all insects or all weeds—we may be thankful for that! Just a few examples will serve to illustrate the kind of subtle ecological problems that have emerged.

The use of herbicides to control weeds in cereal crops can be associated directly with the increasing prevalence of wild oats as the major cereal weed. The reason is fairly simple. Unlike insecticides, herbicides must be selective. It is not possible to use on a growing crop a substance that kills all plants. That being so, the first weeds to be attacked were those which bore the least resemblance to the crop plant. In the case of cereals, the broad-leaved weeds were sprayed first. As they were brought under control it became possible for a new group of weeds to dominate as their more successful competitors were removed. So that group was attacked and Britain lost many of the wild flowers that so delighted the Romantic and Georgian poets. As Aldous Huxley said, "You are destroying the subject matter of half of English poetry." As group after group is removed, those that remain resemble the crop plant more and more closely, so cereals come to be infested widely and regularly with the one weed it is exceedingly difficult to control chemically.

The problem was compounded by other changes that took place in farming. Probably the most significant was the replacement of separate binding and threshing machines by the combine harvester. Whereas at one time the crop was gathered and taken to one part of the farm to be threshed, nowadays the combine threshes as it goes. Threshing removes many weed seeds from the grain and drops them onto the ground. Traditionally they were deposited all together, but now they are scattered across the field and left to germinate with the next crop.

It can be argued that the situation is worse today than it was when the range of weed species was greater, for although they might reduce yields somewhat, at least they did not contaminate the harvested grain and reduce its value, as do wild oats. Even the proportion of the crop that used to be lost is of less significance than it may appear superficially, since it is known now that herbicides themselves reduce crop yields.

There are three possible solutions. The first is to do what some farmers are doing at present, which is to remove the wild oats by hand, either by pulling them out one at a time, or by using a contact herbicide carried in a back-pack which feeds to an impregnated glove worn by the operator. Obviously, this is a labor-intensive operation the cost of which should be set against the "saving" achieved by the herbicides. The second way is to give up the growing of cereals and admit defeat. In all probability a number of years spent growing other crops would make it possible to reduce wild oats to manageable proportions. The third solution may be to reduce, or stop altogether, herbicide applications. This would allow the old weed patterns to reestablish themselves which would reduce the relative importance of wild oats and create an overall situation it is easier to control.

Insecticides have caused more dramatic imbalances. Because they are nonspecific they kill many nontarget species, most of which are beneficial. It has been estimated that in the world as a whole, there are no more than 3,000 insect species, or 0.1 per cent of the total, which represent a hazard to man by competing with him for food or by acting as a vector in the transmission of disease.[1] Certain of them are pollinators and play a critical role in the production of some crops. Their populations have been reduced considerably from time to time. The insect most observed in this connection, of course, is the bee, because it is domesticated. A hive of bees is a valuable piece of property and beekeepers often arrange with fruit growers to place their hives in orchards at blossom time. The bees pollinate the trees and the blossom imparts its flavor to the honey. It is a truly symbiotic relationship, but the beekeeper is less than charmed when the grower's efforts at pest control put him out of business. Fruit growers on the whole are now alert to this danger.

Other nontarget species are actually allies of the farmer, engaged in controlling the populations of those insects the farmer regards as pests.

Dr. J.P. Dempster and Mr. T.H. Coaker, entomologists employed by the British Government's Nature Conservancy, set out to investigate the effect of DDT on pests. They chose as their pest the caterpillar of the small white butterfly *(Pieris rapae)* on brussels sprouts and they found that after spraying the survival of the pest improved considerably.

The caterpillars lived among the foliage of the plant and while many of them were killed by the spray, those that survived could be attacked only by predators that were able to cross the sprayed plant. The pests, meanwhile, continued to graze the unsprayed leaf that had been missed, or they were naturally resistant to DDT. Many of the enemies of the caterpillar live on the ground during the day and climb the plant stem at night. Not only did they have to cross the poisonous plant to reach their prey, but they suffered from surplus DDT that dripped down to them during spraying. The predators living among the foliage were affected as were the pests, the effect on them being increased by their greater mobility: those that were sheltered and so survived by avoiding contact with the DDT during spraying must nevertheless cross the poisoned plant to feed.

This meant that although the pest population was reduced, the survivors had fewer enemies and so a higher proportion of them reached maturity.

This was not the end of the story, for while the crop was harvested and replaced, the following season, with fresh, un-sprayed plants, sufficient residues of DDT persisted in the soil to keep down the predator population. Once the grower had spray-ed, it was therefore certain that he would have to spray again if he planned to grow a crop susceptible to attack by *P. rapae*. DDT, far from solving the problem, merely served to aggravate it and to extend it from one season to at least a second.[2]

Insect resistance to insecticides is well documented. The common housefly is totally resistant to DDT and most of the organo-chlorines, and there are at least 250 species of farm pests so resistant that insecticides can no longer be used against them.[3] In July, 1970, Professor James R. Busvine, of the London School of Hygiene and Tropical Medicine, told an FAO Working Party on Pest Resistance, of which he is a member, that he knew of 600 cases of resistance, involving more than 100 different species, among insect disease vectors.[4]

The development of resistance is illustrated very clearly by the story of the Queensland cattle ticks.[5] The tick, *Boohilus annulatus microplus* is not native to Australia and so, in common with most introduced pests, it has no natural enemies in Australia although it does have them in the countries of Africa and Asia from which it originates. It was introduced in the 1870s and by about 1900 it had spread all over the north and coastal areas of Queensland. Around 1920, farmers began dipping their cattle to control the pest, using arsenite of soda. It was only partly successful and as soon as DDT became available they changed to it. In the mid-1960s, DDT had to be abandoned because residues were surviving to contaminate produce and as other countries introduced controls over levels of residual DDT, exports were threatened. The farmers changed to organophosphate compounds which tend to be more toxic than the organochlorines, but less persistent. In about 1966 the tick problem became much worse. Farmers began to dip more regularly and for the first time ticks were encountered during the winter. By 1968, strains of tick appeared that were definitely resistant to the "tickicides." From then on the battle became frantic. The state government embarked on an ambitious program of tick clearance to create quarantine areas, the insecticides being paid for by the state. The campaign had some effect, but it was terminated in September, 1969, probably because it was proving too expensive. During the program, farmers had to dip every 17 days with solutions at double strength. The ticks thrived on the treatment and a point was reached where a dose adequate to produce a reasonable tick kill would also have killed the cattle. At the most intensive, farmers were dipping every 10 days, at double strength. More and more different products were introduced and tried, but the speed with which the insects acquired resistance to each new poison was almost uncanny. An average kill was around 40 per cent.

Insecticides have been known to create pests where none previously existed. The red spider mite (which is neither red nor a spider) caused growers no trouble at all until its more vigorous competitors were eliminated by spraying programs. Now it is a particularly common and virulent pest of certain fruits and vegetables and it is resistant to most, if not all, insecticides.

Had the Queensland tick problem been recognized in time, it is probable that it could have been brought under control by importing the ticks' enemies, the mynah bird or the cattle egret, birds which pick ticks from the hides of cattle in India and Africa. Alternatively farmers could have begun a phased changeover to tick-resistant livestock.

The red spider mite has also proved susceptible to biological methods of control. Most of the research on it has been performed by the British Glasshouse Crops Research Institute (GCRI). Insect pests are likely to be a more serious problem in glasshouses than out of doors. There are several reasons for this. Crops are grown much more intensively indoors than out of doors, and in a large commercial operation, plants are sown in one house, transplanted into a second and then, perhaps, into a third, so that plants at the same stage of development are moved together. Should they acquire a pest, it is easy for the pest to be transported all over the unit within a short space of time, and completely unobserved. Once inside the glasshouse, the pest finds an unlimited food supply and its population can expand unchecked, since unless a predator happens to be present already, it is impossible for one to enter. Moreover, those pest insects which swarm and migrate find themselves trapped, so that their population density may well exceed greatly that which normally would cause them to thin themselves by spreading. Because crops grown under glass are more expensive, and therefore more valuable, than outdoor crops, growers tended to use their sprays frequently, heavily and thoroughly. The inevitable result was that the insects acquired resistance more quickly than they could have hoped to do outside.

The GCRI discovered another mite which eats red spider mite and which has a feature most unusual in a predator—it breeds more rapidly than its prey. A technique was developed for the protection of several crops, but especially of cucumbers, which are normally grown commercially under glass in Britain. The grower is advised to plant his crop and then, a number of days later, to introduce the pest. The pest is allowed to proliferate and then the predator is introduced. The predator multiplies more rapidly than the pest, so overtaking it very quickly and both populations fall, leaving behind the small number of survivors of both populations. Any red spider mite arriving from outside will be dealt with very quickly. From this point on the only further control necessary is the periodic reintroduction of the pest to ensure that the predator population does not starve.

The secret of the technique lies in its timing. It is only during the later stages of its development, when the fruit is forming, that the plant suffers economic damage from pests and so the early proliferation of the pest on the young plants may look alarming to the grower, but in fact it does little real damage and does not reduce his yield.

Because they have been compelled to discover for themselves viable alternatives to insecticides, the GCRI now practices biological control against most insect pests and although it must still use fumigants and some fungicides, all of these are tested for their effects on insects before they can be used. Occasional accidents, such as one which occurred at a weekend, when the more experienced staff were away and a young assistant permitted an untested fungicide to be used and killed large numbers of insect predators, cause much distress.

Yet biological and cultural controls of insects, in one form or another, have been practiced regularly by organic farmers for many years. Until recently they were regarded as reactionary and deliberately obtuse to reject the offer of such a self-evident boon to the farmer as insecticides.

In the United States it is possible for farmers to consult entomologists who will advise them on biological methods of pest control and who will sell them colonies of predators.

It is known, too, that as well as depressing yields of the following crop, herbicides can cause pest problems if they are used too efficiently, simply by removing alternative sources of food for insects. This observation, made by many farmers, led Dempster and Coaker, who investigated the effect of DDT on *P. rapae,* to experiment with a more diverse kind of cultivation. Although the presence of weeds among a crop may reduce pest infestation, the competition from the weeds themselves usually more than offsets any advantage. Dempster and Coaker sought, therefore, for a eed crop that would help control the pests yet which would not depress the crop. Again, they chose as their crop brussels sprouts and the pests with which they were most concerned were, again, *P. rapae* and also the cabbage aphid, *Brevicoryne brassicae*. They planted an experimental plot of sprouts undersown with Kersey white clover, at 22.5 kg per ha. On the control plots all weeds were removed by hoeing and the surrounds were sprayed with herbicides. The clover failed to keep down some perennial weeds, such as creeping thistle, and these were cut by hand. They found that the same number of eggs were laid on both plots, but that fewer larvae survived on the clover plots. The *P. rapae* caterpillar causes no significant damage until after its third instar and by that time its numbers were well under control due to the activities of arthropod predators, especially the ground beetle, *Harpalus rufipes* and a harvest spider, *Phalangium opilio*. At harvest, the yield from the clover plots was higher. The experiment was repeated later at the University of Cambridge Experimental Farm, using red instead of white clover. On this occasion the effect on pests was even more marked but yields were nevertheless reduced because the clover developed ahead of the crop and, to some extent, choked it.

Small, family farmers usually have smaller fields than the huge conglomerate operations. Because of this the diversity of plant, and therefore insect, species tends to be greater. This is due partly to the likelihood that a different crop is being grown in each field or group of fields and partly to the fact that smaller fields mean a greater area of headland and field edge. These small areas of relatively wild land permit stable populations to become established and, on balance, it is likely that they contribute to pest control. There are many factors at work and such research as has been done on the benefits of hedges in pest control are rather inconclusive, since while it is true that predator populations may exist on the edge of a field, pest populations may multiply there and migrate into the crop, while incoming migrant pests may be carried by air currents clear of the hedge and some distance into the field, out of the range of the predators in the hedge. In spite of this, the observations of some organic farmers would suggest that a diverse and healthy headland and hedgerow community can help the grower.

Sam Mayall, who is probably Britain's best known and most respected organic farmer, said that once he was walking round his farm with a party of government scientific advisers. When they came to a field of beans, being grown for cattle feed, they found on the headland what looked like the beginning of a serious infestation of black fly aphids. The scientists advised Mr. Mayall to spray quickly and thoroughly. Even had he been prepared to do so, which he was not, it would have been difficult, for the crop was well advanced and was standing too high to be reached conveniently by a tractor-drawn sprayer. When next he visited the field, he found the pests had not migrated into the crop. There had been a "population explosion" of ladybirds (ladybugs) and the aphis problem had solved itself—nature had solved it. Had he sprayed he would probably have made things worse. He has said, too, that in his fields of cereals it is possible to find specimens of all the common fungal diseases of cereals, and yet these never develop into epidemics and never reduce his yields, which are consistently higher than those of many of his neighbors and well above the average for his region. Perhaps the plants are healthier and more resistant, perhaps there are other organisms at work within his field ecosystem which control the fungi, he does not know, but he is a good farmer and he is careful not to overreact.

As we said earlier, overreaction causes more problems than it solves and the farmer who believes he has a pest or a weed when in fact he does not, and who sprays against it, may be providing himself with a new pest or a new weed. It is rather like making a self-fulfilling prophecy. Tell the world that the dollar is about to collapse and maybe it will. Attack an imaginary pest and you may find yourself with a real one.

At the other end of the scale there is Arthur Hollins. He is an organic dairy farmer whose cattle stay out of doors all year round. He is unique among organic farmers in that he makes no compost at all and yet the fertility of his fields appears to be increasing. His secret lies in the extremely complex mixtures of grasses and herbs he sows in his pastures. His theory is that there is no nutrient present in his soil that he cannot reach with the root of a plant, which will take it up and, when the plant dies down, leave it on the surface. He can show two paddocks, side by side, managed and stocked in the same way for many years. One is clearly improving while the other is declining. The only difference between them is that the one which is declining has in it two varieties of grass, while the one which is improving has 27. He is able to support about one cow per acre, which is the average stocking rate for Britain. Obviously, he can have no weed problem, since weeds are part of his crop and he regards them as very important.

It is weeds that worry organic farmers far more than insect pests. While there is much they can do by way of careful and timely cultivations, they are trapped inside the overall economic system that dominates modern farming. Herbicides serve only to save labor. They do what human beings used to do, and, as we have seen, they are more clumsy and less efficient than a skilled worker. The records from the Soil Association's experimental farms show that over the years the sections which were never sprayed with a herbicide suffered less from weeds than did those that were sprayed. An experienced agricultural journalist once complimented the farm director on the cleanliness of one of his fields before he learned that no herbicide had every been used on it. Yet weeds do occur and herbicides have played their part in displacing agricultural workers. The organic farmer finds it no easier to find farm workers and no easier to pay for them than does his competitor and while he would prefer to employ human workers rather than rely on the inferior energy substitute the chemical companies would sell to him, in practice it may be difficult. Yet there are solutions and the organic alternative exists for weed control as it does for every other operation.

The organic solution to weed problems is mainly mechanical, part biological. The use of herbicides, as we have seen, tends to alter weed patterns and to encourage those weeds most closely related to the crop and those that are most difficult to control. On the Rodale organic farm at Maxatawny, Pennsylvania, the use of 2,4-D by the previous owners eliminated most of the weeds, leaving little competition for giant foxtail. This is a particularly noxious weed which spreads rapidly. Not only does it compete with crop plants for light, water and nutrients, but it secretes a phytotoxic substance that inhibits the development of maize. Its prevalence seems to be due entirely to the previous use of herbicides. The first part of the control program worked out by the Rodales is to encourage the reappearance of the original weed patterns that were destroyed by spraying. This may sound paradoxical, but it is not. Since the giant foxtail was no serious problem when there was a wide range of weed species, it is likely that the other weeds, which are easier to deal with and less damaging, will help suppress it. Combined with this, the Rodales have suggested eliminating the row crops which favor the weed. It cannot compete with solid-stand crops, such as legumes or grass. Where row crops must be retained, the answer is mechanical cultivation. They have found a rotary hoe useful when bad weather prevents the use of heavier or slower equipment and the weed is developing ahead of the crop.

The fact is that although herbicides were introduced in order to save labor, a similar result might have been achieved less harmfully by mechanizing the operations involved in weed control. To a large extent this has been done and it is possible for organic farmers, most of the time, to maintain adequate control by the skilled use of machines and implements that are easily obtainable. On balance, organic farmers probably suffer less from weed problems than do farmers who have used herbicides regularly for a long time. Indeed, it is often the first serious signs of failure of herbicides that persuade farmers to convert to organic methods, a failure exacerbated now by the development of herbicide resistance among some weeds.

It is insecticides and herbicides that have been used most extensively and for the longest period and so it is the limits to their effectiveness that have appeared first. However, there is no reason to suppose that similar limits will not appear in the case of fungicides. All that is needed is for more fungicides to be used.

However, the susceptibility of a plant to attack by an insect pest, or a fungal disease, or its relative robustness in withstanding competition from weeds depends on more than the position of the plant in the farm rotation and on more than the use or avoidance of pesticides. In part it is inherited and there are intensive breeding programs under way in most of the industrial countries to develop new, resistant varieties of crop plant. Such programs can and do make a contribution, and of course resistant varieties are acceptable to organic farmers provided they produce crops that are the nutritional equals of older varieties, but the technique is subject to even more immediate limitations than the use of pesticides. Breeding for resistance to one disease or insect may involve breeding for heightened susceptibility to others, even to others that were not previously considered a problem. The severe outbreak of southern corn leaf blight in the United States in 1970 was caused by *Helminosporium maydis,* a common endemic fungus that until that year had never caused serious crop losses. The success of certain corn (maize) varieties had led to their being grown very widely and when the outbreak occurred it was realized, too late, that the entire national crop was at risk. Nor can resistance be maintained, for fungi mutate frequently and unpredictably and return in new, virulent forms, while insects may change their habits to allow them to eat plants that have been so bred as to be unattractive to them.

Susceptibility and robustness are due, too, to the general health of the plant, which is largely determined by the condition of the soil in which it is grown. Plants grown in a poor, depleted or severely unbalanced soil will be more likely to succumb to an attack by pests or disease and will be more likely to be choked by weeds.

There are just a few basic rules that farmers may follow to enable them to contain pests, weeds and diseases on their farms within acceptable economic bounds. A well-planned crop rotation is the first step. Continuous arable and especially cereal cropping is not just bad organic farming, it is bad farming of any kind. To an organic farmer it is anathema. Timely cultivations followed, if necessary, by mechanical weeding, will control weeds provided that the soil is in good condition, with adequate humus, so providing the crop plants with a plentiful and well-balanced supply of nutrients. Pests may be controlled by allowing weeds to grow, provided the weeds do not grow ahead of the crop. This will create a diversity of plant species which will permit the establishment of a diverse insect population. If further control is required, a biological technique may be available and, if not, there is a range of acceptable organic insecticides that are short-lived and nonpolluting. Some are physical in their effect, such as diatomaceous earth; others are based on naturally occurring substances derived from plants, such as the pyrethrins, derris, rotenone and nicotine. Disease may be controlled by maintaining a variety of species of crop plants and, so far as is possible, a number of varieties of each species, so as to minimize losses should an outbreak occur. Mercury-based seed dressings, which are highly dangerous to humans and to wildlife, have been found to be unnecessary by organic farmers. Rotational farming prevents the build-up of weeds, pests and diseases specific to a particular crop, from one year to the next.

It cannot be said too often that an organic farmer cares first and foremost about his soil. One of his prime aims is to improve its structure and fertility until it reaches the highest level possible and then to maintain it at that level. Each season, in each field, he grows his crop, but generally it is a different crop from that which he grew in the previous season. This involves him in no additional work. The land must be prepared, the crop sown, tended and harvested, no matter in which field it is grown. He relies mainly on the natural forces present on his farm to protect his crops for him. He works in such a way as to encourage them to do so. The energy he supplies is partly his own but by choosing to establish a system which, ecologically, is a little closer to the natural, stable climax than that of the large-scale farmer, less energy is required to maintain it. He is farming with nature, rather than in spite of it. In the long run his system is bound to prove the better because it is based on sound biological and ecological principles. It is more truly scientific than industrial, chemical, "scientific" farming.

As arable farming has intensified and industrialized itself, so has livestock husbandry. First chickens, then hogs, veal calves, beef and dairy cattle have been subjected to a form of "industrial processing" that regards them as no more than protein factories. Indeed, the industry has been nicknamed "factory farming" and the first protests against the inhumanities it involved were aroused by a book called *Animal Machines*.[6] That was in 1964 and the concern felt by the general public related only to the welfare of the livestock. Public fears were countered by assurances that the animals and birds were, in fact, well cared for, since they gained weight satisfactorily and no creature under the kind of acute stress that these creatures supposedly suffered would gain weight. Not many people were reassured and it was pointed out by some of the leading ethologists that such an assumption was based on gross ignorance of animal behavior. It does not apply even to humans. It is quite common for men to put on weight while they are in prison, but no one suggests seriously that they enjoy being imprisoned or that given a free choice that is the life thy would select for themselves.

An official committee was established to investigate conditions in intensive livestock units and to produce recommendations that could be used as the basis for legislation to safeguard farm animals. Under British law, farm animals are not protected by the same laws regarding cruelty that apply to laboratory animals or to domestic pets.[7] Some steps have been taken to lay down minimal space requirements and to require the diets of livestock to reach barest adequacy, but the recommendations of the government committee have never been implemented because of the strenuous opposition to them from the intensive livestock industry itself. These recommendations suggested that, for example, "an animal's diet should be such as to maintain it in full health and vigor,"[8] "that adequate illumination should be available for proper routine inspection of the animals"[9] and that birds should have room to spread their wings and animals should have room to turn round in their pens. That such basic requirements have never been accepted by the industry as a whole is an indication of the mental attitude of many rich and powerful farming corporations, although, in fairness, it should be pointed out that many individual farmers do apply the recommendations voluntarily.

Indoor intensive rearing is rather different from the outdoor feed lot that has become so common in the United States. It is curious that the two industries, both aiming to produce the maximum profit in the shortest possible time at no matter what cost to the animal itself should have moved so far apart and that the farmer should not have adopted the cheaper feed lot system. It is possible that a move in this direction may be about to begin. Some researchers have observed that cattle appear to be able to survive out of doors, even when it rains, and that since it would be cheaper to build intensive units without roofs this might be a desirable course to pursue. It may sound incredible—it *is* almost incredible—but it has been reported in the farming press as a matter of some interest. This is the distance agriculture has moved from the simple understanding of the plants and animals that are the raw material of farming that was taken for granted less than a generation ago.

Under some farming conditions, battery hens are often confined three to a cage, with one-half square foot of space each, and cases have been recorded of seven birds being packed into a cage. In most cages at least one bird was dead and in one cage one of the birds had killed the other six. They are kept in semi-darkness, as are birds being raised for the table, each of which may have little more than one-half square foot in which to move about on its deep litter. Many of the birds are almost blind. Veal calves are confined in pens or crates five feet by two feet in dim light. By the age of three weeks they are unable to turn round. They are fed a milk diet with just enough iron to prevent them from flagging and they have no roughage, which is a biological necessity for a ruminant. Some 15 per cent of all the sows in Britain spend their whole lives confined in crates which prevent them from turning round and hogs being raised for pork or bacon have little more than room to lie down.[10]

It is obvious that no one who understands the organic alternative could approve of such treatment of animals and that no organic farmer could practice it. However, the physiological stress to which the animals are subjected is only one of many arguments against "factory farming."

Housed in expensive buildings, fed automatically, illuminated however dimly by artificial lighting, the stocking density is dictated to the operator—it would be obscene to call him a farmer—by

Farmers who may invest as much as $120 per acre to spray their land with pesticides are finding that these chemicals are creating more problems than they are solving. Many insect pests have become immune to the very pesticides that were created to wipe them out. And farmers are forced to use heavier applications of these chemicals each year, severely upsetting the natural pest population, which includes the beneficial insects as well as the pests. Because of the increasing ineffectiveness, costs, and dangers of pesticides, some farmers are now consulting companies who specialize in rational insect control programs. These specialists are recommending many of the same natural methods used by organic farmers, like introducing beneficial ladybugs and trichogramma wasps to prey on insect pests.

the need to service the capital investment. Having therefore achieved the highest stocking density possible, the system has based itself on the most extreme form of monoculture. It is subject to all the limitations of any monoculture only whereas in certain circumstances the growing of one crop on the same land repeatedly may represent a departure from the situation to be found in nature, the close proximity of one plant to another may not. In livestock monoculture, on the other hand, such close proximity itself is highly unnatural and so the imbalance is greater and so will be the adverse effects. The natural response to population densities of this order is to reduce the population, if not by starvation or fighting, then by disease. Fighting is a problem that can be regulated, at least to some extent, by restricting animals still more severely, but disease presents a more serious risk.

An apparent short-term "solution" to the disease problem was found as the result of a happy coincidence. Just as livestock operators were becoming aware of the risks they ran, the pharmaceutical industry was reaching the limits for the expansion of the drug market. Surpluses of antibiotics were being produced and the livestock market opened up at exactly the right time. However, the therapeutic use of drugs was not enough to save the livestock units. By the time an outbreak of disease had been diagnosed, no treatment could prevent it from spreading rapidly through the unit and eliminating a large part of the stock. So drugs were used prophylactically. They were added, in small doses, to feed and then it was found that in these doses they stimulated growth. The addition of drugs to animal feeds became standard practice. At the same time, insecticides were used heavily inside livestock units to control insects which act as disease vectors.

We have discussed at some length the disadvantages of the use of insecticides. Drugs produce a similar result among bacterial populations to that which insecticides do among insect populations. Beneficial organisms are killed as well as harmful ones and it is only a question of time before bacterial strains appear that are resistant to the drugs and efficiency decreases.

When doctors administer a drug to a human patient, or when veterinarians administer a drug to aid an animal in overcoming disease, the doses given are sufficiently large to achieve an optimum kill. That is to say, sufficient pathogens must be killed to permit the beneficial organisms present in the body to complete the job. A course of antibiotics is seldom repeated on the same patient until sufficient time has elapsed for all of the possible side effects to have disappeared and, of course, it may not be repeated at all. Thus the development of resistant strains will take a long time and may not occur at all. This is the sensible use of drugs and had insecticides been used in this way most of their side effects might not have occurred.

The prophylactic use of drugs is a very different matter, and much more akin to the repeated, routine use of insecticides. The doses are smaller than therapeutic doses and they are given regularly. The use of drugs as growth stimulants is the most dangerous of all, because they are administered each time the animal is fed in doses that are too small to have any significant effect on microorganisms. They are almost guaranteed to produce resistant strains.

The first drug-resistant bacteria were identified in Japan in 1959 and from then the phenomenon proceeded rapidly.[11] The organisms were *Shigella* strains, which cause dysentery, and they were resistant to streptomycin, chloramphenicol, tetracycline and sulfonamide. Some were resistant to only one drug, but most were resistant to all of them. Resistant strains of *E. coli* were discovered in Britain and Germany a few years later and by the mid-1960s drug resistant bacteria were a common phenomenon.

However, these were resistant drugs associated with human disease. Resistant strains had been identified in animal units as early as 1956 in Britain.[12] It is likely that the resistant organisms which had caused illness in humans originated in livestock units and were transported to their human hosts via animal products.

In 1965 there was a severe outbreak of salmonellosis among men in Britain and at the end of 1967, 10 infants died from gastroenteritis in Tees-side. The deaths were attributed to *E. coli*, 0128, which was resistant to eight of the 11 antibiotics tested in the Middlesbrough public health laboratory.[13] There was no conclusive evidence of animal origin but the case was suspicious, to say the least.

The resistance of a particular strain to more than one drug was known as multiple resistance and in 1967 it was found that resistance may be transferred from one strain to another by infection. If a nonresistant strain comes into contact with a resistant strain, in a short space of time both strains are resistant. This came to be called multiple-multiple resistance and the implications were serious. A resistant strain of a relatively harmless bacterium may infect a strain that is highly virulent. At about the same time it was learned that resistant strains grow more rapidly than nonresistant strains.[14] The most probable explanation for this is that the drug which did not kill the resistant strain did kill nonresistant competitors, so removing a constraint on their multiplication.

The picture that emerges is very like that which we associate with pesticides in general and insecticides in particular. Ecological balances are disturbed, so inhibiting natural control mechanisms, and populations acquire resistance. In the case of drugs the situation deteriorated more rapidly and more dramatically than with insecticides, which is what our theory would lead us to expect. Bacterial populations are much larger than insect populations and they breed far more rapidly.

Of course, there is one profound difference. No matter how serious a pest an insect may become, insects that attack plant crops are not parasitic to man. They do not attack man directly. Even those insect parasites which vector disease do not cause the disease themselves—it is the bacteria they carry and which infect the wound made by the insect. Bacteria, on the other hand, create problems in livestock precisely because they are parasitic to mammals and many of them parasitize man as readily as they do any other animal host. If the drugs that we have developed to combat infectious disease lose their effectiveness, human populations are immediately at risk.

You might suppose that governments, in their wisdom, would forbid farming practices which exposed consumers to this kind of health hazard. They have been advised to do so repeatedly by bacteriologists, doctors and veterinarians. Yet they have not done so, at least not completely. There are regulations in the United States which prohibit the use of a list of drugs for prophylactic purposes. Nowhere has there been an outright ban on the prophylactic use of all drugs that are used in the treatment of human disease. There has not even been an outright ban on the use of drugs as growth stimulants.

More recently, attention has focused on the use of hormones as growth stimulants. The fear here is that traces of the hormone may remain in animal products to disturb hormonal balances in the body of the consumer. Again, there is no code of practice that is accepted internationally.

Intensive livestock rearing is cruel to the animals and birds it regards as "protein factories" and it presents an immediate hazard to the health of the consumer. It also causes severe local pollution, since its wastes cannot be disposed of in the usual way, by composting and returning to the land. The intensive unit, or feed lot, permits a stocking density far in excess of that which could be achieved under an extensive, or free-range, system of management. This is its attraction and it is this departure from sound farming practice that causes its disadvantages. Because animals housed indoors utilize less bedding straw—often they stand on a slatted floor to permit urine and fecal matter to drop into a pit below—what used to be called farmyard manure is now called slurry. It is liquid. It can be used on the land directly, with some care, but it is more difficult and far less pleasant to handle than manure. Often it is sprayed onto land as a disposal method but since the land area available to the operator seldom bears any relationship to the size of the herd or flock that could be quoted as a stocking density, the effect is to sterilize the land, poisoning plants and soil populations and causing rapid oxidation and depletion of humus. Land that has been treated in this way for a number of years is likely to be left a desert for a long time after the livestock unit has gone.

The quality of produce derived from such units is highly suspect. Evidence is beginning to be available that shows both eggs[15] and meat[16] are inferior to those produced on free range.

Yet many people continue to believe that we must accept cruelty, the dehumanization of workers—for this is closely associated with routine, casual cruelty to animals—a threat to our own health, environmental pollution and inferior produce because there is no other way in which food can be produced in quantities adequate for feeding the world.

Figures which are quoted would seem to bear this out. In Britain, for example, in 1866 there was a total of 4.786 million cattle. By 1945 the figure had risen to 8.697 million but in 1970 there were 12.581 million. There were 2.478 million pigs in 1866, 1.903 million in 1945 and 8.088 million in 1970. In 1866 there were about 12.4 million poultry (including ducks, geese and turkeys). In 1945 there were 40.615 million and in 1970, 143.43 million. These were the animals and birds that were subjected to intensification. Sheep, which have come under considerable criticism from agricultural scientists for their steadfast refusal to grow more than two udders, which makes them grossly inefficient, have not moved into intensive units, but into marginal hill land. Consequently, their numbers have not increased significantly for a century. In 1866 there were 22.048 million; in 1945, 19.496 million; and in 1970, 26.080 million.[17] This represents a major increase apparently as a result of intensification and since intensification began in the 1950s and accelerated through the 1960s, much of the increase has taken place very recently. Surely, such an increase in production must represent an increase in efficiency?

A closer examination of the figures will show this is not so. Since arable production has not increased per acre since the early 1960s, increased livestock production has been achieved in the only two ways possible. First, grassland no longer required for grazing has been ploughed up and sown to barley for use as animal feed. This has led to a situation in which barley dominates British arable farming in a way no single crop has ever done before. It has necessitated monocultural cereal growing and this, in turn, has led to all the imbalances we associate with monocultures. At the same time, the loss of grass from those farms which still practice rotational farming has been a major contributing factor in the decline of soil structure. This, in real terms, is inefficient. Further increases have been made possible by the simple expedient of increasing imports of feedingstuffs. At present roughly half of the feedingstuffs used in Britain are imported. It is not particularly efficient to increase food production by increasing the import of food! The very high capital cost of intensive units and the high energy subsidies required to transport materials into and away from them, and to operate them, add to the waste with which they are fast becoming synonymous.

So far as is known, no one has calculated the efficiency of intensive livestock rearing in terms of energy use, but it would seem to be the most inefficient farming system man has ever devised.

As to its contribution to world hunger, trade in livestock products takes place almost exclusively among the rich nations, with the single exception of the import of feedingstuffs, especially fishmeal imported from Africa and Latin America to feed the United States and, even more, Western Europe. Indians do not buy broiler chickens from us because they cannot afford to do so. Insofar as it is relevant to the world food situation at all, the contribution of "factory farming" of livestock is to take food from the poor and hungry and give it to the rich and well fed, wasting about 90 per cent of it in the process.

The organic alternative has no place for such livestock systems and organic farmers will have nothing to do with them. Organic farming is based on the mixed farm, where livestock play a proper, integrated part in the rotation. They are raised and fed out of doors on grass and herb mixtures that have not been chemically treated and when they have to be moved indoors they are fed from organically grown produce. Their superior health, more serene temperament and the superior quality of their produce is obvious and indisputable.

chapter VII

ORGANICALLY RAISED MEAT

Often, concern over the apparent failure of chemical farming systems motivates a farmer to convert to organic farming. Even with such motivation, the farmer is a courageous man, for he may well be rejecting much of what he was taught in college and most of the advice he receives from those anxious to sell him their chemical products.

Sam Mayall called in his veterinarian to advise him about his dairy cattle. They were becoming increasingly infertile. The veterinarian told him there was nothing that could be done since the cause of this infertility was unknown. However, if it was any comfort, all Mr. Mayall's neighbors were suffering from the same problem and all of them were receiving the same advice. That was in 1948 and Mr. Mayall, who even then was widely respected as a farmer, thought long and hard. He concluded that his animals might be suffering some dietary deficiency resulting from the way their food was grown. He decided to abandon the use of chemical fertilizers, and so he became an organic farmer. The fertility of his herd improved. Today such a story would not be news. In 1948, however, it suggested very radical thinking, but it was radical thinking that paid. Mr. Mayall is now a very successful, prosperous farmer. His neighbors still have fertility problems!

The 600 acre farm is on the Shropshire-Staffordshire border, and the Mayalls have been there since 1923. The present acreage was reached in 1947 when they bought the farm which adjoined them. They have found that since they changed to organic methods costs have not risen in proportion to output, and output is now higher than the regional average. Mr. Mayall employs rather more labor than do his neighbors, but the labor cost is less than the cost of fertilizers, pesticides and imported feed for which he has no use. He has a pedigree herd of Ayrshire cows and heifers numbering more than 200 and raises hogs to bacon weight and fattens lambs on the grass during the winter. He grows 90 acres of wheat, 70 of barley, and oats and beans for feed. His cows yield more than 1,000 gallons each, with a calving index of 384 days and a total solids content of 13.3 per cent. More than 20 bacon pigs leave the farm each week, as well as 250 lambs in the spring. Each year he raises about 100 heifers for herd replacements and for sale. His wheat yields about 35 cwts per acre, against a national average of 32.2, but the figure is, perhaps, less important than the fact that the yield is maintained, whereas yields dependent on heavy fertilizer inputs are subject to quite wide fluctuation. So while Mr. Mayall does not achieve the exceptionally high yields of 2 tons an acre or more that his competitors sometimes harvest from one or two fields, neither does he suffer their reverses. His wheat is milled on the premises for sale as compost-grown, stone-ground flour under the brand name "Pimhill," which is to be seen in most British health-food shops. In addition to his flour, he mills and sells oats as breakfast cereals.

While it operated its own experimental farms, the Soil Association obtained its breeding bulls from an organic farmer who won the Cardiganshire Dairy Herd Competition for all breeds for seven years in succession. Ms. Dinah Williams has a 250 acre farm overlooking the Dovey Estuary, near Aberystwyth. The land is wet and exposed and only 171 acres can be ploughed, but her Guernsey cows yield an average of more than 900 gallons of milk and sometimes more than 1,000. The 900 gallon average has been maintained for close to 20 years. The stocking rate is roughly one cow per acre, with 20 acres of rape. Ms. Williams imports some feed concentrates, but her profit per cow and per acre is more than adequate. She believes that a cow that yields well for two lactations is less profitable than one that starts with a lower yield but goes on producing for 12 or more lactations—and she has proved that this is possible with cows that have produced 14 or more calves.

The economics of organic farming, or, indeed, of farming of any kind, is closely related to the price paid for the land; and it is not surprising to find that the farmers who can afford to aim for the highest standard of husbandry, rather than the highest, fastest yield, have owned their land for a long time, or have inherited it. Michael Scully, who farms 410 acres in Sangamon County, Illinois, inherited land that had been in his family for more than a century, part of a large family estate. One hundred acres are pasture, 80 are hay and the remainder is cropped in a rotation of corn (maize), oats and clover for three years, followed by sudan grass, corn, oats, clover for the next five years. Mr. Scully raises 150 beef cattle and 500 hogs and sells around 200 of the hogs and 120 head of beef cattle a year. Manure is composted and applied at 8 tons an acre in the second and third year of the rotation, together with 2,000 gallons of diluted liquid manure in the third year.

It is the quality of his meat that impresses his customers. The butchers who handle his produce tell him that from 25 to 75 per cent of the livers of nonorganically produced beef animals are condemned, while he has had only one liver condemned, and that was the result of an injury. The livers and hearts of his animals are smaller, and Mr. Scully believes that the enlarged livers and hearts of animals fed commercial rations are a product of the stress to their metabolisms. His animals are skinned more easily, which suggests there may be changes in the structure of sub-cutaneous fatty tissue in commercially raised animals, and the beef is much redder than commercially fed beef, which is streaked with gray.

There are unsolicited testimonials, too, for his pork. The owner of a trucking firm told him he had not eaten pork for five years before he first encountered Mr. Scully's pork. Now he is able to eat pork chops up to five times a week!

Very gradually, organically produced meat is finding its way into American freezers. The shortage of clean, wholesome meat, free from antibiotics and hormones, is almost pulling a large-scale organically grown meat business into existence. At the same time it may be creating an exciting new form of direct marketing which could offer a valuable alternative to many North American housewives who have access only to the meat stocked in their local supermarket.

Dave and Jan Hays have been selling beef direct from their small farm in Nelson, California. The farm represents the fulfillment of a dream for Dave, who describes himself as "a country boy from the city of Los Angeles" who used to long to earn his living by producing food. Although the farm is not yet completely organically run, this is the aim. In pursuit of it he may be establishing a production-marketing system that could change significantly the current patterns, which favor the feed lot.

He has worked out two concepts which could make a practical proposition of the production of organically raised beef on a large scale. The first concept he calls the "locked-in herd." This means the rancher sets aside a particular herd on a basis of its health and history, and from that point on it is raised organically, detailed daily records being kept not only of the animals themselves, but also of the management of the fields on which they are grazed. Adoption of this technique would make it possible for large or small ranchers to produce beef organically, even though they were not running an organic farm in the conventional sense. It also marks a departure from the customary feed lot, or intensive indoor unit, in which animals are moved arbitrarily according to the stage they have reached in their development. Each move causes a weight loss and produces tensions within the animal. Young, immature animals are trucked into feed lots, confined in small grassless pens, stimulated and "exploded" into 1,000- or 1,100-pound chunks of plastic meat, then rushed to slaughter. It is a tense situation in which tense animals are made still more tense, and the system might collapse in a series of widespread epidemics of disease were not antibiotics used liberally and prophylactically.

The "locked-in" herd remains in one environment in the fields or hills until the animals reach a mature 850 lbs. Then they are driven down from the hills. If the cattle drive is conducted with care, animals lose very little weight.

The second concept is called "free-choice feeding," which contrasts with the feed lots' forced-feeding. The animals are given a free choice between the fresh grass in their field and a bin containing dry feed. This allows them to follow their natural inclination. The feed lot, on the other hand, forces animals to consume a controlled diet while confining it to restrict mobility in order to realize the most rapid rate of weight gain possible. Cattle may put on weight at a rate of 4 or 4.5 lbs a day.

The "free-choice" animals, on the other hand, gain no more than 2 lbs a day. This is deliberate. The aim is to finish the animals slowly and naturally. More rapid weight gains increase the deposition of fat, which is known to be harmful to the health of the consumer. An animal will be finished out at between 900 and 950 lbs, and the goal will be to produce meat high in protein and with no more than a quarter of an inch of fat.

"Free-choice" feeding maintains an open, natural environment. Consequently, healthy animals are not easily susceptible to disease. Ample room allows adequate exercise and rest. Drugs are used only to treat actual disease or infection.

Eventually, Mr. and Mrs. Hays hope to be able to buy all the feed they need from neighboring farmers who will have been contracted to grow crops organically. Beyond that, it is possible that the whole "free-choice" concept may become established permanently on a nearby 7,000 acre ranch.

The meat will be sold and delivered direct to consumers. Mr. Hays has bought the first of what he hopes will be a fleet of refrigerated trucks that will deliver to the doorstep. He believes he has a formula for success in wholesome food, profitable production and efficient selling.

Clearly the production of organically raised meat is on the increase. Standards, definitions and rules of conduct for organic growers have always started by laying down procedures for cereal and vegetable growers, and it is this side of the organic market that has developed first. This is logical, of course, since there is little point in discussing standards for livestock husbandry until the methods by which animal feed and pastures are to be grown and managed have been agreed and until an adequate supply of organically grown feed is available. Yet, in a sense it is paradoxical. Upland sheep farming in some countries is more or less organic in that no agrochemicals are used and animals are pastured out of doors all year round; in spite of the impact of the intensive beef ("barley beef") industry in Britain, Scotch beef is still recognized as having the highest quality on the market, and Scotch beef is raised outdoors on grass. Of course, the grass receives chemical fertilizers, but the change to an organic system of management would involve no very radical departure from present practice. As fertilizer prices increase and production costs approach levels that reflect their real efficiency, is it likely that Scotch beef producers will change to organic methods? It is at least possible.

In France there are moves to coordinate organic food production and marketing, led by Nature et Progrés, a progressive, dynamic voluntary organization that has been established for several years and that has close contacts both with Rodale Press in the United States and with the Soil Association in Britain.

One of France's leading organic farmers raises beef. His name is Christian de Monbrison. His love of the land is instinctive, but it is also reinforced by the fact that the area in which he farms has been called Monbrison since the thirteenth century and his own land has been in the family since 1732. He bought his 400 acre farm from a cousin in 1961. When he began, he followed the advice of neighbors, but soon he began to question it. He had considerable business experience and he realized that his enterprise was tied to fluctuating fertilizer and pesticide prices and to skilled labor he might not be able to replace should his workers leave. So he decided to change to organic methods, aiming to achieve a high degree of self-sufficiency and so reduce his imports. He suspected, too, that were he to abandon agrochemicals the health of his animals might improve as well as his soil. He read books on the subject, visited Britain and met Sam Mayall. He came to realize that organic husbandry is based on facts that are well known but buried beneath loads of statistics. He realized, too, that government departments are usually wrong. He began thinking for himself and he worked out a system of organic grassland management to suit his own conditions. His farming story was told by Robert Waller, former editor of the *Journal of the Soil*

Association in the May, 1973, issue of Rodale's *Organic Gardening and Farming.*

"Christian started his pedigree herd of Blonde d'Aquitaine— the regional breed—in 1964, and now has 130 head which are kept out of doors all the year round. Although the summers are warm, the winters can be severe with snow. The cattle are raised on 400 acres of grass interspersed with woodland. Calving takes place outdoors without assistance on a maternity pasture, as he calls it, near the corral and barn, where help, if needed, may easily be given, although this is seldom necessary. (Throughout France the Blonde d'Aquitaine are known for their easy calving.) He has 64 cows in calf during the year—and aims at 80—plus three service bulls. The followers consist of 20 heifers and 30 female and 60 bull calves. These animals account for 67.24 tons on the hoof. Animals have been exported to New Zealand, Canada and Britain; this breed provided the first export of semen from France to the US in 1971 (45,000 vials)."

" 'I have found that by sticking to these principles of selection (of his stock) the animals can put on weight as fast as any kept under confined conditions, where they cost more to keep, for they have to be provided with litter and require the constant attentions of the herdsman. Mine are out of doors looking after themselves. They feed on grass which doesn't have to be brought to them. Consequently they are much healthier and I don't have many vets' bills. Confined conditions spread and cause disease.'

"In addition to his 400 acres of grazing pastures, 200 to 260 tons of hay from the farm are cut and baled. Sixty to 100 tons of straw are brought in from local farmers who would otherwise burn it. This is used for outdoor bedding that is turned into compost in the summer.

" 'The average growth rate of my stocks is 2.2 lbs per day, and in some cases up to 4.4 lbs or even 5.5 lbs. . . . The weight of a calf at birth is around 117 lbs for a male and 163 lbs for a female. . . . I have no trouble with sterility or abortion or bloat. Seventy-five per cent of the cows calve within 12 months after their last calving, although the Blonde d'Aquitaine is known to have the longest pregnancy of all beef cattle.'

"Christian always keeps every acre covered in the summer so

that it is not baked by the heat. He has given up using the plough, becoming a great believer in subsoiling and using the chisel plough."

Many farmers have doubts about the advisability of ploughing, and with good management a permanent pasture practically renews itself each year with no damage to the soil structure. He sows as many as 16 different grasses and herbs, which root to various depths and provide the animals with a diet which changes through the year to provide for variations in the animals' natural requirements. The system is best for the soil and it is best for the animals. War-time experiences in Israel and Egypt have taught him two "tricks" which help to conserve moisture.

"When his cattle poach the land in wet weather so that the water lies in the hoof marks, it looks terrible. But when the land dries out, there are two or three inches of soil above the water. This makes it possible to take in the tractor and discs, and by moving at a fast rate, break up the surface without bogging down. The water, of course, keeps the seedbed moist. Another of Christian's tricks is to throw down hay and let the animals sow the seeds in with their treading, thus reseeding the pasture."

It sounds like bad farming, but it is not. He survives droughts.

He recognized that cattle are woodland animals. Their natural habitat is woodland clearings and grassland at the edge of woodland. He regrets cutting down one of his woods on the advice of an "expert," who told him that land would earn more money as pasture. As woodland it offered shelter for his stock during the heat of the day. As pasture he can graze it only at night in summer, because of the heat during the daytime. All farm animals are still related very closely to their wild ancestors, and left to themselves they will revert very quickly. Even today, with hybrid stock bred specifically for indoor rearing, there is no species that would not survive quite happily in the wild.[1] Cattle that are completely wild are difficult to manage. They are shy of man (with reason!) and the cows hide their calves. If such problems are to be avoided, Christian de Monbrison believes they must have frequent contact with humans.

"All the meadows are mown in the early spring and the grass

left on the land to feed the microorganisms and bacteria and worms that do all the fertilization. A month or two later, the second cut is done. The meadows are subsequently grazed by cattle or cut yet again. During the rest period, the pastures are fertilized with rock phosphate and seaweed, so that afterwards they will carry more beasts per acre.

"By all these means Christian can fatten 150 beasts on 350 acres over two and a half years helped only by one herdsman. As a preventive measure against disease, he never buys replacements for his herd; he turns over every possibility of improving fertility by imagining how nature works."

A major influence in the agricultural philosophy of Christian de Monbrison was a wise French agronomist, André Voisin.[2] Voisin was an authority on grassland management and opposed the use of nitrogen fertilizers on grass, basing his fertilizer programs on the use of calcified seaweed, Maërl Émeraude, which comes from Brittany. The use of seaweed as a fertilizer is the basis of one of the main schools of organic farming in France, where it is called "le méthode Lemaire-Boucher."

Ron Lee is an English farmer, from Somerset, who has also been influenced by Voisin. He uses seaweed, too, but what makes his system especially interesting and productive is his "controlled grazing."

His 6 to 8 acre fields are divided into a total of 35 paddocks, each of one to one and a half acres in area. His entire dairy herd, of between 100 and 110 Jerseys, is packed into one paddock, where it spends 24 hours before being moved on. After 24 hours of this treatment, the pasture has been grazed very closely and the area is heavily dunged. Even so, once the herd has gone, the paddock is topped over with a forage harvester. The object is to remove those tufts of grass the cattle have left. This grass has not been eaten because the animals found it unpalatable, and if it were left, it would have grown larger and, perhaps, more extensive by the time the paddock was grazed again, so that over a period of time the pasture would deteriorate.

Many farmers advise against the heavy dunging of grass, believing that this makes the pasture sour. Mr. Lee—and his cattle

—dispute this, but it may be that the phenomenon does occur when cattle are fed diets grown with chemical fertilizers, which alter the composition of the dung. Even at this dunging rate, Mr. Lee still adds the manure from the yards where his cattle spend the winter. He winters the cattle in yards because, in his view, the land needs to rest, just as does a human. The climate is mild and wet, though, and his land is wet in winter.

Only 21 of the paddocks are used in any one growing season— they are bounded by electric fences—the remainder being used to grow hay and for young stock. A prominent feature of the farm is the stack of 100 to 120 tons of straw he buys in.

In a county of lush, green pastures Mr. Lee's pastures come close to being the lushest and greenest of all. His system permits him to stock 160 nonpedigree Jerseys on only 65 acres and his milkers yield 800 gallons per year each, which works out at 1,000 gallons per paddock, with a solids-not-fat content of more than 9 per cent and a butterfat content of 5.5 per cent.

Every organic farmer tries to farm with nature, rather than in spite of it. Tom Lasater, who ranches 27,000 acres near Colorado Springs, Colorado, takes this much further than most. He has allowed his land to revert to its natural climax, and his only modifications to the natural ecosystem are his substitution of a herd of 1,100 Beefmaster cattle (a strain bred from the Braman, Shorthorn and Hereford) for the bison that grazed the land before the first European settlers arrived and his use of windmills to provide water for the animals.

He refuses to kill "pests"—even rattlesnakes—and he has more than sufficient evidence to justify this policy.

When he first arrived on the ranch, in 1950, it was infested with rabbits. He was advised to "do something," but he did nothing. Soon he noticed that the number of coyotes was increasing and soon after that the number of rabbits began to decrease. Eventually the surplus coyotes moved on and he was left with a balanced rabbit-coyote population.

On another occasion he found an infestation of porcupines among his cottonwood trees. He noticed that they were eating dead wood—in effect, pruning the trees—and he left them. In time they moved away. "Then I heard from some other people around here that they're losing their cottonwoods. The trees are dying all over the place. But not on *my* place. So I asked some of these people if there had been a lot of porcupines around, and they said there had been until they shot them out of the trees. Cowboys will shoot at anything. They've been doing their best to ruin this land for a hundred years."[3]

On the one occasion when he did take advice, he killed a town of prairie dogs, which had occupied a 20 acre range. A year later a new foreman who knew nothing of the incident asked him why the grass was better there than anywhere else on the ranch and he realized that far from damaging the grass, the prairie dogs had been helping to improve it. He hoped more prairie dogs would arrive, and when they did not, he imported some.

In fact, Mr. Lasater likes "to sit back and let nature do the work. She's a hell of a lot smarter at it than we are."

He culls his herd, taking out calves that show a distrust of men and removing cows that do not produce calves. "Either a cow puts a calf in the weaning pen or she goes to the sales pen herself." This may sound ruthless, but it only imitates natural selection which would remove infertile strains. He is less worried about the performance of his bulls. The less fertile strains are eliminated naturally over a few generations.

The bulls are extremely tame—"They eat out of our hand on the range"—and none of them is castrated. "A cattleman gets an immediate 10 per cent yield on beef production by stopping unwarranted castration."

The breeding season lasts for 65 days in the middle of the winter and heifers are bred at between 12½ and 14 months of age.

The ranch's problems, and it has problems, are all that is imported. The water level is falling in the aquifers that surface on the ranch because of the activity of irrigators and the city of Denver.

One of the most "natural" organic farms in Britain must be that of Arthur Hollins. He has farmed all his life. He left school early to take over the 150 acre holding when his father died. At that time the farm was very run down and the early years were hard. Chemical fertilizers had been used heavily; the soil structure was poor and yields were low. Mr. Hollins, too, used as much fertilizer as he could afford to buy, and it was not until he applied the muck from his pigs and the spent compost from his mushroom enterprise to the land and yields began to rise that he realized that all his soil lacked was organic matter. He had to study compost in order to make a success of mushroom growing and he met Sir George Stapledon and Sir Albert Howard, two pioneers of organic farming. Nevertheless, he was not convinced that the improvements he could observe were the product of the compost and muck alone. He began to rotavate his pastures instead of ploughing them, and the results convinced him that the plough was to be regarded as an enemy. Today he does not plough at all.

He and his wife opened a country club and this gave them the idea of converting their dairy herd to Jerseys in order to provide guests with cream and milk. Nowadays he exports clotted cream to Devon—the home of clotted cream! It was the demand stimulated by these guests that led the Hollins to open a stall at the weekly market in Crewe, some miles away, selling clotted cream, soft cheeses and, later, yogurt. Slowly the dairy business grew at the expense of other crops.

He began to experiment with leaving his animals out of doors during the winter, first with half of the herd and later with the full herd. He experimented with chemical fertilizers as well, fertilizing half of a crop of kale and leaving the other half unfertilized. Although the fertilized crop appeared to yield more, he found that the animals grazed both crops at the same rate, covering the same area before lying down, satisfied, and that there was no difference in milk yield from either section of the herd. The increased yield produced by the fertilizers had no cash value.

When the cattle were out of doors all year round, he noticed they preferred to graze the edges of the fields, close to the hedges. This suggested that there might be a deficiency in the center of the field, so he sowed a greater variety of grasses and herbs and began to experiment with deep-rooting herbs. Already Mr. Hollins suspected that it might be possible to by-pass the composting process. He reasoned that nature makes no compost and yet is able to sustain high rates of fertility. The traditional way for the farmer to utilize this fact was by rotational farming, with a fallow. However, he could not afford a fallow, so he wondered whether he could achieve a similar result if he grew a sufficiently diverse mixture of plants. He undersowed his arable crops, but this was not enough. All plants create their own microenvironment, encouraging their own soil micropopulations, so the greater the variety of plants, the greater will be the variety of microorganisms and this leads to a biological balance, which means a nutrient balance. His own soil should be acid, but in spite of the fact that he has not limed it for many years, it is not. So, over the years, "weeds" have become his main crop. One of his simpler pasture mixtures consists of Tetrone ryegrass (4 lbs), Perennial ryegrass S23 (6 lbs), Perennial ryegrass S24 (6 lbs), Cocksfoot S26 (3 lbs), Cocksfoot S143 (3 lbs), Timothy S51 (1 lb), Meadow fescue S215 (1 lb), Perennial ryegrass Taptoe (4 lbs), Perennial ryegrass Petra (4 lbs), White clover S100 (2 lbs), Chicory (2 lbs), Sheep's parsley (1 lb), Kidney vetch (1 lb), yarrow (⅛ lb) and Nettle (1 lb).

Having rejected ploughing, he is not altogether satisfied with rotavating, which cuts too deeply into the soil. All the energy in the soil, which is put there by plants and by microorganisms, is in the top four or five inches. The activity of soil microorganisms can be stimulated, but ploughing or, indeed, too much disturbance from any cause tends to suppress it by bringing to the surface soil from a lower, less biologically active level. He believes that land that is ploughed will need fertilizers, and he has designed a machine that slices off the top inch or two of soil, leaving a firm bed containing many roots that will rot down to leave channels for drainage and aeration. The sliced layer is passed back into the blades of a cutter that mix, chop and aerate it and then throw it out behind to lie on the surface as a seed bed. It contains plants, soil and microorganisms, which are the ingredients of a compost heap. As it ferments it heats very slightly, so stimulating the germination of the seeds he sows.

Now all the cattle are out of doors all year round and although there is indoor accommodation for calving, usually this takes place out of doors as well.

Production is not very high, the Jerseys, stocked at about one

cow per acre, yielding an average of only 600 gallons of milk
a year with a butterfat content of about 5 per cent, which is above
the national average but not so high as that achieved by some
organic farmers. The animals are grazed on a paddock system,
for the purpose of which the farm is divided into four sections for
spring grazing, summer silage, summer grazing, and autumn and
winter grazing. A small dressing of 8 cwts per acre of fish and
bone manure is applied in August to the autumn and winter
grazing area and in January to the spring grazing. Full spring
grazing begins four weeks earlier than on most farms in the area.

Arthur Hollins is a happy, fulfilled man. He likes to meet his
customers, and between 4,000 and 5,000 people visit his farm
every year, arriving by the busload. Visitors are met, shown a little
of the farm, talked to, given a meal, and leave feeling that they
have seen a little of organic farming in action. This contact with
his customers is important. In an autobiography he has written
but so far not published, Mr. Hollins has described his farm: "a
meeting place, courtship and marriage of a man to his environment, his soil, plants, animals, weather and finally his products
and his customers." This is what organic farming is all about.

Of course, Arthur Hollins is a dairy farmer. His range of "Fordhall" products—butter, soft cheeses, yogurts, and coleslaws on a
yogurt base—are to be found not only in health food shops, but
in leading grocery stores throughout Britain, and in recent years
he has been exporting to Europe. There are organic beef farmers,
but the market for organically grown beef is very small. Organically
produced beef, lamb and pork are sold through normal marketing
channels and lose their identity. Probably this is due to a lack of
marketing organization, rather than a lack of consumer demand.

Hugh Corley raises beef and lamb organically on 185 acres
of flat, mostly heavy clay land in the Upper Thames Valley. He is
known to those interested in organic farming as the author of a
book called *Organic Farming*, published by Faber and Faber in
1957.

Mr. Corley has farmed his present holding since 1938 and
began with beef. During the war he was compelled to plough up
the grass he had sown and change to dairying, but in 1963 he

returned to producing grass-fed beef. That was a time when milk prices were low and many dairy farmers were being forced to change. In fact, beef is a much more suitable enterprise for his heavy land than relatively more intensive milk. He grows no arable crops at all, and this is a matter for regret, since he believes some arable land is good for any farm. As it is, he has to buy in barley and cake, when he would prefer to feed his animals dredge corn with beans added for calves and dairy cattle.

He thinks the grass would be improved if he could in-winter all the cattle, but this would call for more capital invested in buildings and it would increase the amount of manure that had to be handled.

He raises Hereford/Angus crosses, of which he has more than 120 and sells around 60 each year as fat cattle. There are six nurse cows which mother about 10 calves each lactation. The calves were brought in, and in the early years they were fed milk substitutes, but when he lost a quarter of them with a variety of ailments, including scour, he determined that in the future they would be suckled. The calves are housed indoors and weaned onto a mixture of barley, Milkiwey—a by-product of butter and cheese production, then condensed—calf nuts, vitamin and mineral supplements and hay. There are some disease problems among the calves, which Mr. Corley attributes to a lack of resistance to local infections among animals that are stocked relatively densely. He is trying to overcome the problem by moving the calves on to grass one day in every three so as to immunize them. Spring calves may spend their first year indoors.

The fields are from 6 to 30 acres in area, and grazing is controlled only by the size of the field. Most of the growing stock winter out of doors, fed on hay, straw and a supplement which stimulates the rumen flora. He has a yard than can hold 60 animals but it is little used.

He also has about 70 ewes of a range of breeds that have been crossed among themselves in as many ways as possible, but they are a better flock than they sound. The best lambs can weigh as much as 90 lbs at 10 weeks and at slaughter they may weigh nearly 140 lbs. Mr. Corley's aim is to so crossbreed that he establishes a flock that normally produces, and raises, triplets. If he is able to achieve this he is certain to make agricultural history. The lambs sell at a good price and Mr. Corley attributes his good results to the high quality and careful management of the pasture and to the standard of shepherding. The shepherd is Mrs. Corley, the only employee!

The manure is stacked in windrows where it decomposes until there is time to spread it. It is not turned. Compost, properly made, would produce better results, but there is simply not time to make it. Fish and poultry offal are mixed into the manure to help it rot, and he has noticed that the sheep grazing fields treated with fish-offal manure consumed fewer mineral supplements. From time to time he applies rock phosphate and, very occasionally, potash. There are 26 trace minerals in this mixture and Mr. Corley believes trace minerals are of great importance. All the animals have access to mineral licks.

In addition to the cattle and sheep there are bees producing honey, as well as hens, ducks and geese providing a few eggs— all sold to local health-food stores or at the farm gate.

In an interview in the *Journal of the Soil Association* some years ago, Mr. Corley was asked why he farmed organically:

"Mostly a matter of principle. Because it pleases me to produce highest quality food. It's being in tune with nature rather than fighting against it. Once you start using chemicals it is rather like the slippery slope; you use one and you need another and so on, then drugs, then more chemicals. I don't see how it can pay to use nitrogen when you can get it for nothing by good husbandry."

Michael Thompson raises sheep on 125 acres of rolling land on the southern end of the Yorkshire Woods. The soil is a light loam, overlying chalk with quite a number of stones and flints. All of the farm is cultivable, and he produces cereals, potatoes and sheep in rotation. He employs one man and he makes a profit.

The farm was inherited. It had been in the family for a century; and prior to 1949, when Mr. Thompson took over, it had grown a rotation of clover, wheat or oats, turnips (folded off with sheep) and undersown with barley. Fertilizers had been used in moderate amounts, but Mr. Thompson stopped using them. Today the rotation is temporary grass for one year supporting the sheep, followed by spring oats, winter wheat and then half the area sown to potatoes, part of which may be left fallow, spring wheat, spring barley and the other half sown to three crops of barley with grass undersown to the final barley crop.

About 25 acres of the farm is grazed and it supports 120 ewes with their lambs, as well as providing hay or silage for the winter. In summer the grassland is divided into eight forward-creep grazing paddocks, and after each grazing the paddocks receive 10,-000 gallons per acre of pig slurry. Combined with the close grazing, this treatment produces a very dense sward; there are no scouring problems and the sheep eat the grass right down. During the summer the lambs are given an additional daily ration of a quarter pound of oats, barley and seaweed, the seaweed providing minerals that might be missing from the pasture, which contains no deep rooting herbs. Lambs that eat more than their share are the first to go to market! During the winter they receive silage and waste potatoes, and the expectant ewes are fed oats, barley and seaweed, with beans being introduced later. Mr. Thompson aims for a 200 per cent production rate, but so far he has not succeeded in achieving it.

Weeds are a major problem on the arable crops. Straw is chopped and then cultivated into the surface and he seldom ploughs more than six inches deep. He does use a herbicide but not more than once in a complete rotation—and then only on barley which is sold off for stockfeeding. He is looking for a break crop that would provide an opportunity for better weed control.

His yields of cereals are good and well above national averages—35 cwts an acre of barley, 36 of wheat and 35 of oats—and the crops are disease free, although in fairness he points out that he lives in an area noted for its freedom from crop disease. His potatoes yield 7 to 8 tons an acre, which is about the national average.

He manures with broiler house manure and pig slurry as well as muck from his yards, but none of this is properly composted.

Mr. Thompson feels a need to intensify production and, while he would prefer to continue to farm organically, he is not sure just how he will achieve this.

Organic farming is not necessarily easy. Certainly it is not the

easiest of all ways to farm, but the climate is changing and there
is little doubt that in the next few years the rate of change will
accelerate.

chapter VIII

GETTING TO KNOW ORGANIC FARMERS

Gösta Olsson[1] sells direct to his customers. They stand with their bags ready while he lifts potatoes from his 100 acre farm on the windy Swedish island of Öland in the Baltic. He farms biodynamically, according to principles laid down by Rudolf Steiner. Biodynamic farming is practiced fairly widely in Switzerland, Germany, France, the Netherlands and to some extent in Scandinavia. The system is based on orthodox organic methods with the additional use of special preparations of minerals and manures, applied in extremely small doses. The timing of every farm operation is regulated very precisely.[2]

Mr. Olsson has 60 head of livestock—cows and pigs—and in addition he grows wheat and rye as well as carrots, beets, onions, parsnips, cabbage and brown beans. Yields are slightly lower than they would be were he to use chemical fertilizers, but the quality and flavor of his produce are superior and he receives much better prices than the average farmer.

The soil is sandy with a lot of chalk and the climate is on the dry side, although when it does rain it rains heavily. The soil is examined regularly at a biodynamic laboratory near Stockholm. All recognized biodynamic growers in Europe are entitled to use the trademark "Demeter," which guarantees the origin of their produce, and all registered growers must submit to these regular checks. The laboratory does more than investigate his soil, how-

151

ever. It also advises him on procedures he should follow to correct any deficiencies it detects.

Mr. Olsson has been farming biodynamically since 1950, having started in order to see whether organically grown vegetables would improve his brother's failing health. It was so successful that the brother became a vegetarian! Olsson makes compost from manure, soil and vegetable matter, together with biodynamic preparations, and he claims that he has no pest problems. In an interview with Ruth Nilsson in *Organic Gardening and Farming,* he said, "Since we stopped using artificial fertilizers and started farming biodynamically, we haven't had any parasitic insects on our farm. Previously there were bugs on our carrots, but this organic method has put an end to that problem. The earth is a living organism, and can be compared to a human being. If a person is healthy, he will have a built-in resistance to colds and other diseases. If he isn't healthy, then various diseases can take over. We enrich the soil with a lot of natural compost, and we rotate our crops to keep the nitrogen content high. I once had a few bugs on my cabbage plants, but they disappeared after a short while. The soil had become so rich through proper treatment that it had its own natural resistance to pests."

Mr. Olsson is not the only organic grower who has found that organic methods prevent pest outbreaks. Arthur Bower is retired now, but for many years he had a large market garden near Wisbech in Cambridgeshire. He was growing three-quarters of a million rose plants on 16 or 17 pieces of land, and after five successive plantings there were still no aphis problems. He said, "I know from experience that we shall not get aphis on organic saps. I can grow salads, I can grow strawberries, I can grow roses without aphis. I have had surveys on my land by young undergraduates. One example I can give you. They were taking an aphis count in this particular area; they were trying to survey the different varieties of aphis. They went over a lettuce bed of mine and they could not find a single count. But 36 inches away in the same basic soils—only separated by a single strand of wire—it was infested. Now they were in kicking distance, they were in wind distance and everything; yet there was not one."

Mr. Bower used to make his compost carefully and in his day

he probably knew more about compost making than any other grower in Britain. He should, for he was taught personally by Sir Albert Howard. He met Howard by chance at a demonstration of agricultural machinery in 1941. He had a long conversation with Howard without ever learning his name, and it was not until he tried to follow the advice he had been given and failed that he began to try to trace the gentleman who seemed to know so much about making compost. Howard visited his holding several times to give personal instruction until eventually he was satisfied that Mr. Bower really knew how to make good compost.

Even then, Mr. Bower was not fully convinced of the benefits of using compost rather than uncomposted manure. This came later.

"We had put four dutch beds down to peonies: these were root cuttings. I told the chaps to mulch them with stuff from the compost heap. They wheeled the heap out onto two of the beds and two half beds; then they came up and told me that they'd run out of that stuff. I said, without any thought at all, 'Well that's all right, muck the rest of the beds at the same rate.' " And that was that. Nothing happened until the end of May, early June: the chap in charge there came up and told me one morning that there was something wrong with the peonies and I ought to go down and inspect them. I told him there couldn't be anything wrong with them because there wouldn't be anything showing, not this year—it was a two-year job. But when I went down, the two old beds and two halves were really an interesting sight to see. We'd got a tremendous take of these peonies. And two half beds were showing the odd leaf here and there as I would have expected to be normal. So I told my chap not to worry. That November, one year before the time, we were able to dig up these two full beds and two half beds. . . .Then it dawned on me, looking back, that these two full beds and two half beds had had this fermented compost, whilst the other two half beds had just had ordinary farmyard muck. Not only that, but when we started digging them up the soil was so kindly and friable—and we were really used to this land—that we started new thinking about compost."

No wonder! They conducted trials of their own using compost

against farmyard manure and against commercial fertilizer. The composted land gave a 3 lb Cos lettuce in six weeks, the manure gave a 1 lb 10 oz lettuce in eight weeks and the fertilizer gave a 1 lb 2 oz lettuce in 13 weeks. The weights were not those of any individual lettuce—they were the weight of the entire crop divided by the number of heads—an arithmetical mean and likely to be conservative. Of course, one small trial proves nothing in isolation, but this kind of result was repeated with other crops, and Arthur Bower became a compost grower. Moreover, his peonies sold for double the current price on the depth of their color, their scent and their longevity.

"I've got a halo that has been given me by the Advisory Service—that Arthur Bower could grow in soot. Arthur Bower could grow anywhere, but let me tell you that Arthur Bower can't. It was his fertility that was doing it and nothing else."

He believed that in the lettuce trial the compost had provided nutrient in an available form, which he called "touch fertility." The manure was not immediately available; it had to be broken down by soil microorganisms. The sulphate of ammonia killed off the "touch fertility."

He found that in spite of the very large amounts of compost he was applying, the pH of his soil was rising while free calcium was falling until analyses showed there was none in the soil at all. With the help of the Soil Department at the University of Cambridge he found that the soil pH rose from March to reach a peak in September.

Lee McComb, who has 5,000 citrus fruit trees on 50 acres in Florida, also knows the value of compost, although his composting method is quite different from Arthur Bower's strictly orthodox Indore. Mr. McComb first moved to Florida with the idea of making and selling compost. The compost itself was to be his crop—one he felt could not fail because, with its light soil, Florida needs all the compost it can get. He was proved wrong so he used his stock to grow fruit trees.

Composting materials are many and various and he imports them from far and wide. Cow manure and dried chicken manure he obtains locally; Hybrotite comes from Georgia; Litterlife made from poultry waste comes from Georgia; and Micro-Min, a trace mineral clay, comes from Mississippi. Rock phosphate and dolomite are found naturally in Florida.

Weeds and grass growing profusely between the trees are cut from time to time and allowed to decompose where they fall. The compost is spread before it is completely mature, being thrown from a truck driven through the orchard, at between 5 and 25 lbs for each tree. A more unusual ingredient in his compost is the worm manure he buys locally. Earthworms are fed peat moss, which makes them grow rapidly (factory farming for worms?), and the by-product is worm castings which are sold by the cubic yard.

Applying the compost before it is mature saves labor costs that would be incurred were it turned, and Mr. McComb has no doubts about its efficacy. The material, now spread on the ground in sheets, continues to decay; and although this might damage vegetable crops, it clearly benefits trees. Costs are kept to a minimum so that prices of their produce compare favorably with those for fruit grown with artificials. Each year he sells about 2,000 bushels of oranges and 5,000 bushels of grapefruit.

The fruit is not harvested until it is ripe and colored, usually between the tenth and twenty-fifth of November, and all of it is hand-picked, taken to a packing house, washed in clean water, graded and boxed. No more fruit is picked at any one time than is needed for packing that day or the following day. Most is sold by mail order and some by the roadside.

Like Arthur Bower, Lee McComb finds that insect pests present little problem. "We depend on a natural balance of insects for control. There is evidence that plant food has a profound influence on the insect problem. Our insect problem is not of serious consequence. There is some scarring of fruit by pests, but not enough to reduce food quality. In addition to ladybugs, praying mantis and other friendly insects, we have citrus snails in our groves that tend to keep the fruit and leaves clean."[3] Citrus snails are natives of Florida. They crawl all over the trees, spreading a substance which traps mites. Every effort is made to keep them in the groves, and from time to time stocks are imported.

Like many other organic farmers all over the world, Mr. McComb is very optimistic about the future. "I have never talked to any housewife who was not interested in buying the most nutritious food she can get for her family, if the price isn't too much above market prices. This means that the demand for organically grown fruit and vegetables will grow with the increase of production and economic marketing channels. No one will buy chemically grown produce if they can get the more flavorful organically grown produce."

The first prerequisite is a market. For the professional grower this means more than a vague notion that people are more interested than they were in the quality of the food they eat and that organically grown foods are more likely to be popular today than they would have been a few years ago. It means going out and identifying and studying the market. It means getting to know the prospective customers—or the retailers who serve them—and finding out as precisely as possible what it is that they will buy. In this sense organic food production is no different from any other business. The person who enters it without having made the preliminary studies he would make before setting himself up in any other enterprise deserves to fail, no matter how much he knows about growing. His aim is to serve his customers and so he must find out, from them, how this may be done. If he is successful he can be very successful indeed.

For many years Bob Bonner ran a 5 acre vegetable farm near Oxford. He sold his produce in the covered market in the city of Oxford and, later, through a shop he opened in an Oxford suburb. His profits were highly satisfactory by any standard, and he lived well. The operation was organic "because my father was so successful with (organic methods): he won a wad of awards in shows and he never saw any reason to change. And I don't see any reason to change either."

The holding employs a manager, two men, two boys and two women who work part-time. A quarter of an acre is under heated glass and there are 800 cloches and 500 Dutch lights to produce early salad and vegetable crops. Eight tons of tomatoes are produced from about 2,000 plants, and cucumbers are timed to be ready when the season ends in the Canary Islands, from which Britain imports its early cucumbers. Manuring works on the principle "feed the soil and the plant looks after itself," with composted farmyard manure being applied at 20 to 30 tons to the acre, as well as fish manure. Weeds are controlled by steam sterilization under the glass and beneath polyethylene sheeting on the cloche and Dutch light ground. The small area of open ground is rotavated, which is less expensive than using herbicides.

In an interview with Robert Waller, published in the *Journal of the Soil Association,* he said, "The first essential is to have a reliable market on your doorstep. And it should be on your doorstep because almost the only advantage of the small grower is that he can get his produce to the shops *fresh* and get a premium price for it. The shops that have to buy through a big central depot, even their own, don't have that advantage."

Mr. Bonner did not mark his produce as "organically grown." In part this was because he saw no special advantage in doing so. "Obviously in the shop we sell all the quality produce we can get, whether it's organic or not. It is the only way we can meet the public demand. We sell organic produce side by side with orthodox produce. The public buys it, I think, on its flavor. It sells well: there is no need to label it. The public comes back for it without knowing how it's grown." This was written several years ago and today the public is better informed, but the message for the consumer is clear: if you want organically grown produce, you must ask for it.

Mr. Bonner grew a variety of crops, but not strawberries. There are few commercial growers producing strawberries organically in Britain, and the reason is a cause for continual complaint from organic gardeners. "That's a skeleton in the cupboard of research. The old Royal Sovereign that we used to grow was *improved* by the East Malling Research Station and was announced to the trade generally as a much better strawberry than the traditional type. By improved they mean increased yield per plant. Unfortunately in improving the original strain, the inborn immunity to botrytis was lost and as a result, despite every effort, Royal Sovereign are no longer commercially grown. A few growers persevere with them but they have to be continually sprayed." Nor did he grow plums because of the damage from bullfinches whose numbers appear to have increased since the introduction of the Wild Birds Protection Act!

Glenn Graber has a farm in Hartville, Ohio, ten times the size of Mr. Bonner's. He produces five truckloads a day of organically grown lettuce, radishes and greens in a daily total of 3,500 packages. He employs between 60 to 100 people and owns three refrigerated trucks himself. The others he hires. He has farmed in Ohio since 1955 on dark, rich soil which he believes exceeds by far the 3 per cent humus content required for Rodale certification on that land. He fertilizes it with rock minerals, phosphate and granite dust, kelp and liquid seaweed, as well as with growing cover crops of rye, sweet clover and oats which are ploughed in early in the spring.

Even larger is the Ypsilanti, Michigan, operation of Thomas Vreeland, with two farms of 734 and 414 acres, producing mainly wheat, soybeans and corn (maize). He has been producing food organically since about 1965. "It was rough for the first four years," he said,[4] but "we can make both ends meet as organic producers."

At the other end of the scale, the small farmer sometimes has other sources of income, and this is a pattern we may see developing over the next few years. With land prices rising it becomes increasingly difficult for private individuals to enter farming unless either they are very rich or they have an income large enough to service a heavy debt. This may be possible only by working part-time at another profession. It is what William Jarvis has done. He is a building worker in Pittsburgh but also owns a 62 acre farm, with seven pigs, one cow and one and a half acres of salad and vegetable crops. He aims to become the organic supplier for his local area and to give up the building industry. This, he estimates, will take a few years.

As the strain of urban life becomes more and more acute and as the cost and inconvenience of urban life become more self-evident, the demand to leave the city and become involved in the production of food grows. As it grows, people find more and more original ways of achieving their ambition. There is the commune movement, based on groups of people—generally, but by no means always, young people—who pool everything they have to buy a piece of land on which they live together. There are businessmen retired early who form small consortia to purchase a holding which one of them farms on behalf of the group. There are young people who wander the countryside looking for opportunities to work on the land and to become integrated into rural communities, a new kind of migrant labor force. The people, their views and attitudes on most subjects and the economic arrangements they devise for themselves are as diverse as it is possible for them to be. This is an excellent thing, for ecological stability calls for the greatest possible diversity and there is no reason to suppose that, up to a point at any rate, this does not apply as much to human societies and individuals as it does to plant and animal species. They would find it difficult to meet and talk with one another were it not for the central philosophy they hold in common. They have a profound respect for the land, for the planet on which they live, for the complexity of the astro-geo-bio-social system of which they are part. They seek to understand it, not so that they may have power over it, for they recognize that such power is self-defeating and illusory, but so that they may become one with it, restoring the unity man lost when first he alienated himself from nature and from his own nature in pursuit of mastery. They are interested in organic farming and only in organic farming. They are the organic alternative.

chapter IX

CAN ORGANIC FARMING SOLVE THE ENERGY CRISIS?

Our troubles continue to mount. By the middle of 1973, when this book was written, the United States had experienced the first wave of the energy crisis. Everyone had predicted it, but there had been a "credibility problem." The story is told of the inhabitants of a Polynesian island who encountered Europeans for the first time. They were going about their business as usual when there appeared in their bay a sailing ship so much larger than their own canoes or than any vessel they had ever seen or could imagine that their minds refused to accept it. To the sailors on the ship it seemed that their arrival had gone unnoticed. They put a boat ashore and found they were ignored completely. It was some time before the truth was discerned. Since their ship was incomprehensible, the islanders *did not see it.* Nor did they see the first sailors to step ashore. The energy crisis loomed and was discussed in some circles as early as 1953 and probably earlier, but warnings were ignored because the implications of such warnings were so outside conventional economic and political thinking as to be incomprehensible—and therefore inaudible. Yet it has occurred.

As it happens, most of the warnings were wrong. What had been foretold was an exhaustion of reserves. What happened was a continuing and, to use a fashionable word, exponential rise in demand which exceeded the technological capacity of the energy industry. It was not that the reserves were exhausted, but simply that they could not be extracted and processed fast enough. The first limit to growth has turned out to be technological.

Were the technology to be improved, however, as doubtless it will, the reserves will become exhausted that much more rapidly than was predicted and the warnings may be wrong again. In each case they have erred on the side of caution or, if you are addicted to the high-consumption rapid growth society, of optimism. If and when the United States overcomes this short-term setback, it will be by extracting lower grades of fuels at the expense of the environment and by increasing imports, which may provoke an energy crisis in Europe and Japan more quickly than might have been expected. The forecasters were wrong again.

The year 1973 saw widespread famine in the Indian subcontinent and in Central Africa as a result of prolonged drought. Relief to the stricken areas in Africa began late because, incredibly, the extent of the famine was not known. The first warnings were not received until famine-hungry people from remote areas bordering the Sahara began to cross frontiers into other states. Relief in famine situations depends on world food stocks. For several years prior to 1973, stocks had been dwindling. By 1973 they were at a very low level. The FAO hoped that a good harvest in the United States would replenish them, but early indications are that floods have reduced the North American grain output. Relief will be little more than a token. The world has little food to spare, and the one country that can supply something is short of fuel to transport it. Indeed, 1973 may be the year in which the world learns that no matter what may be the theoretical long-term possibilities for food production, the reality is that there are large areas which cannot be fed now.

The incredulity persists. The suggestion that the industrial way of life as we have come to understand it may not be sustainable for very much longer even in the richest countries of the world is still so radical, so outside the experience and comprehension of a generation of world leaders produced by the social system that is threatened, that it is dismissed with political bromides based on the intellectual contortions of advisers who cannot bear to see their entire value-system challenged. For this is what the ecological crisis does. It challenges society to examine its aims and its values and to pronounce judgment on them. We are told that we have achieved the highest standard of living the world has ever seen. Have we really? Is it not possible that a causal relationship exists between wealth and poverty and is this not an ethical matter? If we seek to hold on to our way of life until the last possible minute may we not do so by widening ever further the gap between rich and poor as the number of the inhabitants of the planet who are able to participate in this life decreases?

It is very probable that this will be the course chosen by our governments, but even if we acquiesce in their decision out of fear of the alternative, eventually, in our children's lifetime, or that of their children, that last possible moment will arrive and the reckoning will have to be paid to the poor of the world whose power then will match our own. Is this a world in which we would choose to live? Is this the future we would choose to build for those who must follow us? For, make no mistake, it is our children who will pay the price.

Do we wish to continue to produce our food in such a way that the land we leave is less fertile than the land we inherited? Do we wish to continue to produce our food in such a way as to endanger the physical well-being of future generations by introducing into *their* environment chemical substances not found in nature, or in quantities and concentrations not found in nature, that may one day be proved harmful, in the only way acceptable to the worst of modern science—in damaged bodies and minds?

The choice is an ethical one, but it is not one to be feared. As we have tried to show, an alternative does exist. There are many who practice it today, and their number increases with every month that passes. It involves no less, and no more, than a subtle change in our viewpoint. Zen Buddhist literature is full of attempts by those who have achieved enlightenment to describe their experience. One monk is reputed to have said that in the beginning he saw only mountains and trees; in the middle they ceased to be mountains and trees; in the end he saw only mountains and trees. But he saw them in a new way. His viewpoint had changed, but that change had liberated him from, among other things, his materialist obsession.

In fact, as society is forced to consider seriously the possible

existence of limits to its own ethic, individuals within it are applying more rigor to their examination of its operations than would
have been acceptable, or even possible, ten years ago. We can
read[1] that a large farm tractor represents an energy subsidy to
farming of 72,600 kilocalories (Kcal) for every hour of its working
life, when we take into account the energy used in its production
as well as the fuel which powers it. We can read that potassium
chloride fertilizer costs 556 Kcal per pound; phosphorus pentoxide,
the soluble constituent of most phosphate fertilizers, costs 760
Kcal per pound; and that liquid ammonia, the basis of many
nitrogen fertilizers, is most expensive of all, at 5,250 Kcal per
pound. These energy subsidies exceed by an order of magnitude
the energy value of food that can be produced by them. Pesticides and herbicides cost even more—11,500 Kcal per pound. We
can read figures such as these and cease even to be surprised by
them. They are what we would expect. All they tell us is that
chemicals and machines are less efficient than men. We knew
that, or we should have. Men are more sophisticated, more obser-
vant, more flexible, more subtle than any machine or chemical
product and so more efficient than they can hope to be in con-
ducting operations involving living systems. And should we be
surprised to discover that nature has devised ways of growing
plants more efficient than our own? We may profit by natural
systems and, within limits, exploit them, but is it not arrogance
to assume we have the skill and power to supplant them?

So we can read the other side of the same story and not find
that surprising either, for, again, it is no less than we would
expect. We can learn that 40,000 acres of lettuce in Imperial
Valley, California, may be condemned because it carries un-
acceptably high residues of a new miracle pesticide, "Monitor-4,"
whose labels read: "Liquid insecticide is an outstanding new
organophosphorus compound which controls a wide range of
pests by contact and local systemic action. It is extremely effective
as a foliar spray against resistant and nonresistant species of
aphids, plus cabbage-looper, beet armyworm, tuberworm and
others. It features excellent clean-up capability and residual protection which, supplemented by local systemic activity, continues

This chicken raising set up is indeed a form of factory farming. Chickens are crammed into rooms so closely they can barely move about. The whole place is lit artificially to increase egg production, and the mash and water are full of antibiotics, yolk coloring material, patented medicines and other drugs. The chickens are de-beaked to prevent pecking; there are no roosters. The birds do little else but eat and lay eggs until production drops off, after which they are killed and sold off to be eaten.

Contrast the picture above with the open, naturally lit environment of the chicken yard shown here, in which hens and roosters run relatively freely. They are bright, alert and healthy, and their eggs are naturally fertile and nutritious.

late into the growing season. Monitor-4 can also be adapted to a highly efficient preventive program for cole crops, head lettuce and potatoes. Its protection has proven valuable in promoting increased yields. Applied as recommended on cotton, Monitor-4 provides effective control of several major pests, including mites and thrips. It is compatible with most commonly used fungicides and is noninjurious to plants when used as directed."[2]

What the manufacturer means, of course, is that it is extremely poisonous and persistent. There is no doubt about its "excellent clean-up capability" provided we can agree about who or what it is that is to be cleaned up. Residue limits were set at 1 part per million (ppm) for lettuce, brussels sprouts, cabbage, cauliflower and broccoli and 0.10 ppm for cotton and potatoes. Actual residues were found on lettuce at 1.5 to 5 ppm, with one alarmed California Department of Agriculture official admitting he had found 5 to 6 ppm in field samples. The chemical has been withdrawn pending further study, but before it was launched on the market it had been studied for three years.

No one in the United States, or anywhere else for that matter, will starve because 5,000 to 10,000 crates of lettuce are lost, but before we dismiss this as just another scare story, let us take one more look at that label. Let us reflect for a moment on the claim that Monitor-4 is "extremely effective." Is it? Will it not aggravate ecological imbalances and will not pests develop resistance to it? Will it not cause at least as many problems as it solves? So far its record is a rather large crop destroyed by the insecticide, some of which might have survived the insects. What about "highly efficient?" In what sense is this product efficient? It claims to have "promoted increased yields." If it cost more than 11,000 Kcal of energy per pound to produce, did its increased yields exceed that input? If they did not, then the claim is meaningless, for it represents one more inefficiency added to an already inefficient system.

One of the aims of the founders of the organic movement was to inform public opinion of the changes that were taking place in the ways food is produced. The movement has been successful, although it has been aided by some spectacular failures of agribusiness. Public opinion is becoming informed very rapidly and farmers themselves are seeking ways to modify farming systems. The ways exist. At present the difficulties are economic, not agricultural. We know how to produce large amounts of food, organically, for a small fraction of the cost we pay at present. It is possible to farm without damaging the landscape, without polluting the environment, without displacing workers and destroying communities. It is possible to do so and at the same time to increase the quality of the food we produce. The economic restriction is due to flaws in the economic theories which govern society as a whole and to the vigorous opposition of those who profit most from the present system.

If farmers are to change, if the organic alternative is to become a reality on a wide scale, then farmers must be informed of what is possible and what help and advice is available to them, and they must be convinced that a market exists for their produce. They will be convinced of this only if consumers demand that their food be produced organically, by asking for organic produce every time they shop. We need a consumer revolution.

Farmers must move closer to their customers in order to be able to provide them with fresh food of the highest quality at the lowest price and in order to be flexible enough to adapt to their local requirements and tastes. Consumers must move closer to farmers in order to appreciate the limitations of food production, to find out what is and is not possible in their area and to discover what might be grown but is not. All of us must alter our viewpoint so that we can learn, together, to live on and with the planet, rather than in spite of it.

In the end, we have no choice.

NOTES and REFERENCES

Chapter I

1 Described by Dr. George W. Irving, Jr., Research Administrator of the United States Department of Agriculture.

2 Roberts Farms Inc.

3 Allaby, M.: 1973, "A new baron?" *Ecologist,* 3, 5, 161.

4 Cleave, T.L., Campbell, G.D., and Painter, N.: 1966, *Diabetes, Coronary Thrombosis and the Saccharine Disease,* John Wright & Sons, Bristol.

5 Recent California study reported by D.P. Van Gorder in *Ill Fares the Land.*

6 Allaby, M.: 1973, "Polychickens," *Ecologist,* 3, 2, 48.

7 For a well-argued attack on reductionism, see Rosjack, Theodore: 1973, *Where The Wasteland Ends,* Faber & Faber Ltd., London.

8 Bear, F.E., et al. (of Rutgers University): 1948, *Soil Sci. Soc. Proc.,* 13.

9 To list all references, particularly recent ones, would be almost impossible and probably pointless. However, for an excellent review of studies and expert opinions describing the hazards of pesticides and demonstrating that the probable hazards were already well known more than 20 years ago, see Biskind, Morton S., MD: 1953, "Public health aspects of the new insecticides," *Amer. Jour. Dig. Dis.,* November.

10 Harris, R. S.: 1960, *Nutritional Evaluation of Food Processing,* Wiley & Sons, New York.

11 Quoted in *TGC Bulletin,* 5, 3, March/April 1973, Cooperative Extension Service, Univ. of Massachusetts at Amherst.

12 Selly, Clifford: 1972, *Ill Fares the Land.* (This is not the same book as Van Gorder's work referred to in reference 5 above.) London, Andre Deutsch.

13 Johnson, D. Gale: 1973, *World Agriculture in Disarray,* Macmillan, London.

14 Johnson: ibid.

15 Based on studies by Michael Allaby undertaken as part of the *Ecologist* sequel to *A Blueprint for Survival.* It will be published during 1973.

Chapter II

1 Allaby, M.: 1970, "The bugs fight back," *Ecologist,* 1, 6.

2 Webster's *Third New International Dictionary of the English Language,* 1962.

3 See Rosjack, T.: *Where the Wasteland Ends,* op. cit.

4 See Allaby, M.: 1972, *Who Will Eat?* Chapter 5, Tom Stacey Ltd., London.

169

5 Observed by Sir Robert McCarrison in 1926.

6 Nath and co-workers.

7 Leong.

8 Hurni: 1941–1945.

9 Antoniani and Monzini: 1950.

10 Parker-Rodes: 1940; Steward and Anderson: 1942.

11 Krasilnikov: 1958.

12 Krasilnikov: 1958; Shovlovski: 1954–1955.

13 Roulet and Schopfer: 1950; Lilly and Leonian: 1935; West and Wilson: 1938–1939.

14 Schmidt and Starkey: 1951.

15 Albrecht: 1958.

Chapter III

1 Testimony presented by Michael Perelman at a hearing before the US Senate Subcommittee on Migratory Labor, San Francisco, California, 11 January 1972.

2 Slesser, Malcolm: 1973, "How many can we feed?" *Ecologist*, 3, 6, 218.

3 Perelman: op. cit.

4 Markhijoni, A. B., and Litchtberg, A. J.: 1972, *Environment*, 14, 5, 15.

5 Slesser: op. cit.

6 Allen, Robert: 1972, "Down with environmentalism," *Ecologist*, 2, 12, 3.

7 Baker, R., and MacGregor, M.: 1971, "The rural iceberg," *Ecologist, 1, 4,* 18.

8 Based on figures taken from *Modern Farming and the Soil*, HMSO, London, 1970. Totals are given region by region.

9 These aspects of the "Green Revolution" were debated at the Second World Food Congress, held by the FAO in The Hague, Netherlands, 1971. The proceedings were published in 1972 by FAO, Rome.

10 Ginsberg, Woodrow L., Director of Research and Public Policy, Centre for Community Change, in testimony presented at a hearing before the US Senate Subcommittee on Migratory Labor, 21 September 1971.

11 See the interview with Geoffrey Rippon by Vanya Walker-Leigh in the *Ecologist,* August 1973.

12 Goldschmidt, Walter: 1947, *As You Sow*, Free Press of Glencoe, Glencoe, Ill.

13 See *Smaller Farmlands Can Yield More*: FAO, Rome, 1969.

14 Gordon, Kermit, of the Budget Bureau. 1966, in Van Gorder, Dan P.: *Ill Fares the Land*, January.

15 Richardson, Len: 1973, "Grain market kaput!" Cover story in *Big Farmer,* February.

16 *World Crisis in Agriculture*: Ambassador College, pp. 19, 21.

17 *Environmental Science and Technology*: 1973. 7, 4, 257.

18 Allaby, M.: 1970, "Where have all the hedges gone?" *Ecologist*, 1, 4, 8.

19 *Modern Farming and the Soil*: 1970, para. 12, HMSO, London.

20 Cooke, G. W.: 1972, *Fertilizing for Maximum Yield*, Crosby Lockwood & Son, London.

21 A Century of Agricultural Statistics: HMSO, London, 1969; and Output and Utilization of Farm Produce in The United Kingdom, Ministry of Agric. Fisheries and Food, London, 1972.

22 Van Den Bosch, Robert: 1971, "The melancholy addiction of Old King Cotton," Natural History, December.

23 See Genetic Vulnerability of Major Crops: 1972, National Academy of Sciences, Washington, DC; see also Allaby, M.: 1973, "Miracle rice and miracle locusts," Ecologist, 3, 5, 180.

24 Harris, Marvin: 1973, "The withering Green Revolution," Natural History, March, pp. 20–22.

25 See also Borgstrom, Georg: The Hungry Planet and Too Many; also Allaby, M.: 1970, "One jump ahead of Malthus," Ecologist, 1, 1, 24, and Allaby, M.: 1972, Who Will Eat? Tom Stacey Ltd., London.

In connection with Chapter III, see also Donahue, Ray L., Our Soils and Their Management, The Interstate Printers and Publishers Inc.; Doane's Facts and Figures for Farmers, Doane's Agricultural Service Inc.; and Selly, Clifford, Ill Fares the Land, Andre Deutsch, London, 1972.

Chapter IV

1 Howard, Sir Albert: An Agricultural Testament, Rodale Press, Inc., 1972.

2 H. R. 14941, presented on 11 May 1972.

3 Golueke, Clarence G.: Composting, a Study of the Process and its Principles, Rodale Press.

4 Spohn, Woerman, Knöfel: 1966, "Glass in compost," Jo. Soil Assoc., 14, 1, 50.

5 The authors of the British A Blueprint for Survival and at least some of the members of the MIT team which produced The Limits to Growth, probably the two major ecological statements to date, all support organic farming specifically and the organic alternative in general, as do many of the world's leading environmentalists.

6 Outlook of Agriculture: 1972. 7, 1. Published by ICI Ltd.

7 Waage, Jonathon: "The parachuting cats of Borneo."

8 See the statement by the Secretary of the Velsicol Corporation, Louis A. McLean, quoted by Graham, Frank: 1970, Since Silent Spring, p. 49, Hamish Hamilton, London.

9 Tinker, Jon: 1973, "Pesticides still the exception on US farms," New Scientist, 58, 531, 848.

10 Pimentel, David: 1973, "Realities of a pesticide ban." Environment, March.

11 Longham, Headley and Edwards, reported in Chapman, Duane: 1973, "An end to chemical farming?" Environment, March.

12 Chapman, Duane: ibid.

13 Krasilnikov, N.A.: Soil Microorganisms and Higher Plants, Academy of Sciences of the USSR, translated and published by The Israel Program for Scientific Translations.

14 Ibid.

15 Murphy, L. S., and Walsh, L. M.: 1971, "Micronutrients play major role in complete fertilizer program," *Irrigation Age,* December.

16 Marx, D. H.: 1972, "Mycorrhizae," *Agricultural Age,* January.

17 *Study Week on Organic Matter and Soil Fertility:* Pontifical Academiae Scientarium Scripto Varia, John Wiley & Sons Inc., New York.

18 Donahue, R. L.: *Soils and Their Management,* Interstate.

19 Reported by Krasilnikov: op. cit.

20 Ibid.

21 Hernando, in *Study Week on Organic Matter and Soil Fertility,* op. cit.

22 Marx, D. H.: op. cit.

23 Bear, F. E., et al.: 1948, *Soil Sci. Soc. Proc.,* 13; and Hopkins and Eisen: "Nutritional and environmental aspects of organically grown food," *Jo. Agric. Food Chem.,* 7.

24 Murphy and Walsh: op. cit.

25 Hernando: op. cit.

26 Mattoni, Rudy: comparative survey for *Organic Gardening and Farming* California Certification, Agri-Science Laboratories.

27 Ibid.; see also Murphy and Walsh: op. cit.

28 Marx, D. H.: op. cit.

29 Macfadyen, A.: 1971, "The Soil and its total metabolism," *Methods of Study in Quantitative Soil Ecology,* Phillipson, J. (ed.), Blackwell Scientific Publications, Oxford.

30 Studies reported by Krasilnikov: op. cit.

31 Belson, K. C.: 1955, Michigan State University Centennial Symposium on Nutrition of Plants and Animals, May.

32 Crawford, Michael and Shelagh: 1972, *The Food We Eat,* Neville Spearman, London.

33 Schuphan, W.: 1965, *Nutritional Values in Crops and Plants: Problems for Producers and Consumers,* Faber & Faber, London.

34 See, for instance, the *Report of the Secretary's Commission on Pesticides and their Relationship to Human Health:* 1969, US Dept. of Health, Education and Welfare, Washington, DC; Mellanby, K.: 1967, *Pesticides and Pollution,* Collins, London; or the SCEP Report: 1970, *Man's Impact on the Global Environment.* MIT Press, Cambridge, Mass.

35 Walters, A. H.: 1970, "Nitrate in soil, plants and animals," *Jo. Soil Assoc.,* 16, 3, 149. Walters produces 189 references.

36 Allaby, M.: 1972, *Who Will Eat?* p. 85, Tom Stacey, Ltd., London.

Chapter V

1 For an excellent history of the agricultural societies, see Hudson, Kenneth: 1973, *Patriotism with Profit,* Hugh Evelyn, London.

2 See Allaby, M.: 1972, *Who Will Eat?* Tom Stacey Ltd., London.

3 *Output and Utilization of Farm Produce in the United Kingdom:* 1972, Min. of Agric. Fisheries and Food.

4 *The State of Food and Agriculture 1972:* FAO, Rome.

5 Papers presented by Fernandes, V. Hernando, and Flaig, W.: 1968, in *Organic Matter and Soil Fertility,* John Wiley & Sons, New York.

6 Krasilnikov, N.A.: 1958, *Soil Microrganisms and Higher Plants*, Academy of Sciences of the USSR, published in English by the US Department of Commerce. Also, studies verifying effective microbial action on various nonwater soluble material, especially granite meal investigation, reported by Pfeiffer, Ehrenfried.

7 *Modern Farming and the Soil*: 1970, Agricultural Advisory Council, para. 11, HMSO, London.

8 *A Century of Agricultural Statistics*: 1968, HMSO, London.

9 Paper presented by Kovda, V. A., in *Organic Matter and Soil Fertility*, op. cit.

10–11 Papers presented by Kovda, V. A., Waksman, S. A., Kononova, M. M., Fernandes, V. Hernando, and Dhar, N. R., in *Organic Matter and Soil Fertility*, op. cit. Results of investigations reported in Krasilnikov, N. A.: op. cit.; Belson, K. C.: 1955, in Michigan State University Centennial Symposium on Nutrition of Plants and Animals, May; USDA Agricultural Information Bulletin No. 299, October 1965; Adams, R. S.: 1971, in paper presented to 5th Annual Conference on Trace Elements in Environment, *Health*, 1 June, Columbia, Missouri; Brown and Smith: 1966, *Agronomy Journal*, 58; Murphy, L. S., and Walsh, L. M.: 1971, "Micronutrients play major role in complete fertilizer program," *Irrigation Age*; Ashmead, Harry: 1970, "Tissue transportation of organic trace minerals," *Jo. Applied Nutrition*, 22, 1 & 2, spring and summer; Van Campen, Darrell: *Trace Elements in Animal and Human Nutrition*, US Plant, Soil and Nutrition Laboratory, Agricultural Research Service, USDA, Ithaca, New York; Albrecht, Wm. A.: *Soil Fertility and Animal Health*, Fred Hahne Printing Co.

12 Paper presented by Fernandes, V. Hernando: *Organic Matter and Soil Fertility*, op. cit.

13 Survey of California soils by Rudi Mattoni, Agri-Science Laboratory for Rodale Press, Inc.

14 Marx, D. H.: 1972, "Mycorrhizae," *Agricultural Age*, January.

15 Kostycheve: 1933, Investigations described briefly in Krasilnikov, op. cit.

16 See Allaby, M.: 1970, "Where have all the hedges gone?" *Ecologist*, 1, 4, 8.

17 Hopkins, et al.: 1966, *American Jo. of Clin. Nut.*, described by M. Schwartz in special report for *Organic Food Marketing*. Also Dr. Homer Hopkins, FDA, Nutritional Research Divn., in memorandum to Dr. Philip L. Harris, Director, Division of Nutrition, 11 June 1965; and Albrecht, Wm. A.: 1971, "Magnesium in the soils of the United States," *Jo. of Applied Nutrition*, 18 November.

18 Hopkins and Eisen: 1959, *Jo. Agric. Food Chem.*, 7; Albrecht, Wm. A.: *Soil Fertility and Animal Health*, op. cit.; USDA Agric. Research Service Report No. 2, "Evaluation of Research in US on Human Nutrition," August 1971; Krasilnikov: op. cit.

19 Results from continuing research at West Virginia University under the direction of Singh and R. L. Reed, reported by Goldstein, Jerome: 1973, "Our 'best' food is no longer best," *Organic Gardening and Farming*.

20 McCarrison, Robert: 1926; Nath and co-workers: 1927–1932; Leong: 1939; Hurni: 1944–1945; Antoniani and Monzini: 1950; Hammer and Maynard: 1942; Sheldon, Blue and Albrecht: 1948. Investigations described in Krasilnikov:

op. cit. Pertains to some results of investigations described by Hopkins in his memorandum to Harris: op. cit. Humic acid implications in Fernandes, V. Hernando: op. cit., and Albrecht, Wm. A.: *Soil Fertility and Animal Health*, op. cit.

21 Results of investigations described in the "Hopkins Memo," op. cit.

22 Bear, E. F.: 1964, *Chemistry of the Soil*, 2nd edition, Reinhold Publishing Co.

23 Murphy, L. S., and Walsh, L. M.: 1971, "Micronutrients play major role in complete fertilizer program," *Irrigation Age*. Also results of continuing research at West Virginia University under Singh and Reed.

24 West Virginia University research, op. cit.; Sheets, O. A., et al.: 1944, S. of Agronomy Res. 68, No. 4, described by M. Schwartz in special report for *Organic Gardening and Farming*, 1972.

25 Bodiphala and Ormrod: 1971, *Can. Inst. Food Technology*, S. 4, No. 1. Also reported by Schwartz.

26 Federal Security Agency: 1951, "Drinking water," *Environment and Health*, p. 6, Public Health Service.

27 Marx, D. H.: op. cit.

28 Parker-Rodes: 1940; Stewart and Anderson: 1942; Krasilnikov: 1958; Shavlovski: 1954–1955; Roulet and Schapfer: 1950; Lilly and Leonian: 1939; West and Wilsen: 1938–1939; Schmidt and Starkey: 1951; Bonner, et al.: 1942; Bonner and Bonner: 1948; Robbins and Bartley: 1922–1938; Robbins and Schmidt: 1939–1945; Bonner: 1942; Ovcharov: 1955; Went, Bonner and Warner: 1938; Crebenskii and Kaplin: 1948; Scheurmann: 1952; Psarev and Veselorskaya: 1947; Matvav and Ovcharov: 1940; Raitkin: 1948; Zakhar'yanys, Gorbacheva and Zglinskaya: 1950; Stephenson: 1951; Schopfer: 1943; Zeding: 1955; Maksimov: 1946; Turetskaya: 1955; Kuba: 1941; Lebedev: 1942; Polyakov: 1949; Carpenter: 1943; Hurni: 1944; Schopfer: 1943; Thomas and Hendricks: 1950. Results of investigations described by Krasilnikov: op. cit.

Chapter VI

1 Williams, Carroll M.: 1967, "Third generation pesticides," *Scientific American*, July.

2 Allaby, M.: 1972, "More weeds, fewer pests?" *Ecologist*, 2, 3, 23.

3 Statement by Frank Wilson of the Sirex Biological Control Unit, quoted in Allaby, M.: 1970, "The bugs fight back," *Ecologist*, 1, 6, 30.

4 Ibid.

5 Told in full in Allaby, M.: 1971, "How Boophilus tricked the tickicides," *Ecologist*, 1, 11, 20.

6 Harrison, Ruth: 1964, *Animal Machines*, Vincent Stuart Ltd., London.

7 Pets are protected by the Protection of Animals Act, 1911, and livestock, inadequately in the view of the reformers, by the Agriculture (Miscellaneous Provisions) Act, 1968, Part 1.

8 Report of the Brambell Committee: 1965, para. 39, HMSO.

9 Ibid, para. 44.

10 Harrison, Ruth: 1970, in *Factory Farming*, Bellerby, J. (ed.), British Assoc. for the Advancement of Science.

11 It appears to have been first reported in English by Watanabe, T.: 1963, *Bacteriological Reviews*, March.

12 Smith, H. W., and Crabb, W. E.: 1957, *Veterinary Record*, 12 January.

13 Nature-Times News Service, *The Times*, London, 28 December.

14 Smith and Crabb: op. cit.

15 *New Scientist*: 12 April 1973.

16 Crawford, Michael and Shelagh: 1972, *The Food We Eat*, Neville Spearman, London.

17 Figures taken from *A Century of Agricultural Statistics*, MAFF, HMSO, 1969, and *Output and Utilization of Farm Produce in the United Kingdom*, MAFF, 1972.

Chapter VII

1 Prof. W. H. Thorpe FRS, one of Britain's leading ethologists, said this in the report of the Brambell Committee, of which he was a member.

2 Voisin died in the late 1960s. During his lifetime several of his books were translated into English and published by Crosby Lockwood and Son, London. Most are now out of print, but the Soil Association, Walnut Tree Manor, Haughley, Stowmarket, Suffolk IP14 3RS, UK, may be able to supply some titles.

3 Quoted in *Organic Gardening and Farming*, February 1973, p. 64.

Chapter VIII

1 Nilsson, Ruth: 1972, "The big organic farm and the man who made it work," *Organic Gardening and Farming*, November.

2 Further information may be obtained from the Biodynamic Agricultural Association, Broome Farm, Clent, Stourbridge, Worcs., UK, or from the Biodynamic Farming and Gardening Assocn., RD1, Stroudsburg, Pa. 18360, USA.

3 Goldman, M. C.: 1971, "Organic oranges—California and Florida style," *Organic Gardening and Farming*, November.

4 In an interview with Maurice Franz: 1972, "Midwest organic farmers make both ends meet," *Organic Gardening and Farming*, September.

Chapter IX

1 Leach, G., and Slesser, M.: 1973, *Energy equivalents of network inputs to food producing processes*. Strathclyde University, Glasgow.

2 O,5-Dimethyl-phosphoro-midthioate, marketed by Chemagro, a division of the Baychem Corporation and Ortho-Chevron Chemical Corporation. Approved by the Environmental Protection Agency on 14 April 1972, Registration No. 2392404 AA, US Patent No. 3.309.266.

INDEX